Reactions to the Market

Reactions to the Market

Small Farmers in the Economic Reshaping of
Nicaragua, Cuba, Russia, and China

Laura J. Enríquez

The Pennsylvania State University Press
University Park, Pennsylvania

Published in cooperation with the Rural Sociological Society.

Library of Congress Cataloging-in-Publication Data
Enríquez, Laura J.
Reactions to the market : small farmers in the economic reshaping of Nicaragua, Cuba, Russia, and China / Laura J. Enriquez.
p.   cm.—(Rural studies series)
Includes bibliographical references and index.
Summary: "Analyzes the reaction of existing and former socialist countries to neoliberalism. Examines economic transitions in agriculture and the reconfiguration of socialism in Russia, China, Nicaragua, and Cuba"—Provided by publisher.
ISBN 978-0-271-03619-9 (cloth : alk. paper)
1. Farms, Small—Former communist countries.
2. Land reform—Former communist countries.
3. Former communist countries—Economic policy.
4. Land reform—Former communist countries.
I. Title.

HD1333.F57E57   2010
338.109171′7—dc22
2009030850

The Pennsylvania State University Press
is a member of the Association of American University
Presses. It is the policy of The Pennsylvania State University
Press to use acid-free paper. Publications on uncoated stock
satisfy the minimum requirements of American National
Standard for Information Sciences—Permanence of Paper
for Printed Library Material, ANSI Z39.48–1992.

CONTENTS

FIGURES, TABLES, AND MAPS

Figures

Tables

Maps

ACKNOWLEDGMENTS

This book would not have been possible without the assistance, patience, and understanding of many people and institutions. The following list includes most of them. However, given its long gestation and the highly politicized environments in which the research was conducted, my memory, as well as my sensitivity to some people's preferences to remain anonymous, may result in the omission of a few of their names. Nonetheless, I very much appreciate all of the support they have provided along the way, while underlining that in no way are they responsible for the book's flaws and failings.

First, and most important, I am greatly indebted to all of the small farmers who allowed me to interview them. Although, undoubtedly, some of them perceived this researcher as a "strange bird" who had drifted in from other climes, most of them were quite forthcoming and willing to answer my perhaps seemingly endless stream of questions. Without their input, clearly, there would have been no book. My hope is that this book will represent a small contribution to the effort to modify political economies and policies so that they are more hospitable to producers of our sustenance. Without them, we will, indeed, not survive.

To reach those small farmers, a number of institutions in both Nicaragua and Cuba played major roles in paving the way to them for me. In Nicaragua, several nongovermental organizations (NGOs) that were working in the regions I chose to study were essential in helping me gain entrée and, in one case, allowing my "team" (which included my research assistant, my infant—in a portable crib, of course—his child care provider, and me) to sleep on the floor of their office while we were conducting the fieldwork. They were the following: INDES, PROTIERRA, and the UNAG in Esquipulas, Matagalpa. In Cuba, the National Association of Small Farmers (ANAP) was my institutional sponsor, as well as key facilitator of access to the countryside and to the people I met and interviewed there. Mavis Álvarez Licea, in particular, deserves very special mention. She gave me the gift of her trust and had confidence, even when I did not, that conducting the fieldwork for this study would be possible. Without her, the Cuban case would never have appeared on these pages. And her colleague, Armando Rama, took the necessary official steps to aid her in this effort. Many other national, provincial, and municipal level ANAP officials and the director and others at the Escuela Nacional de la ANAP also provided me with important ideas, contacts, and hospitality.

The numerous government officials, representatives of NGOs, farmers' organizations, and others whose work affected the small farm sector and who allowed me to interview them in Nicaragua and Cuba also added their *granito de arena* to make this book what it finally became. I am quite indebted to them as well.

In both countries, fellow researchers and others familiar with agrarian issues also became important sources of feedback, offering insights, reassurances, and new questions to my slowly evolving lines of inquiry. In Nicaragua, Adolfo Acevedo Vogl was especially helpful in this sense. In Cuba, Nora Cárdenas Toledo, Armando Nova González, Lucy Martín, Juan Valdés Paz, Gerardo Timossi, and Vilma Hidalgo de los Santos should be mentioned. Nora also shared her wealth of newspaper clippings and informal knowledge with me, while Armando pored over endless tables with me, deciphering them with infinite grace, in the face of my frequent bewilderment about how to interpret them. Nereida de los Santos, Bill Brent, and Jane McManus warmly welcomed me into their lives, providing me with a home away from home during many visits; sadly, Nereida, Bill, and Jane are no longer with us. To all of you, many, many thanks.

Back in the Global North, a number of colleagues facilitated the movement of my ideas from the stage of vague impressions to final editing of this book. At the top of this list is my indefatigable sounding board and dear friend, Michael Burawoy. As has been true for my research in general for the two decades we have worked together, he constantly pushed me to make this project more than I was inclined to do at each stage. Most important, as I was describing to him in, undoubtedly, unbearable detail the devastation I was witnessing being wreaked in the Nicaraguan countryside in the 1990s, he suggested I take a look at what was happening in Russia at that time; and as I talked about the fascinating alternative strategy I found being employed in Cuban agricultural policy, he said I must examine the Chinese strategy of the early reform years. Hence, he was the impetus for this study shifting from a two-country comparison to a four-country comparison, which I firmly believe—and hope—has made it that much richer.

Several other "northern" colleagues who contributed to the development of this project in various forms were Nancy Jurik, Rose Spalding, Kamran Nayeri, Martín Sánchez-Jankowski, Peter Evans, Sandra Smith, Jon Jonakin, Geske Dijkstra, and James Church. Nancy guided me through the process of getting the manuscript in form to send off to potential publishers, approaching that particular "market," and, finally, on the decision to go with Penn State University Press—a decision I have no doubt was the correct one. Thank you so very much, Nancy. Rose made herself available at crucial junctures to provide essential input to this effort as well. Kamran, Martín, Peter, Jon, and Geske offered insightful comments on pieces of or the whole "product." Librarian James Church happily assisted me in tracking down inequality data for pre-1990 Nicaragua, which was no mean feat. And Sandra accompanied me on more walks around Berkeley than I can place a number on, listening to me go over the trials and tribulations of the research, the writing, and getting it "out the door." Now that it *is* out the door, I'm hoping that she will continue to go on walks with me despite the effort that it will represent to find new topics of conversation.

Although I may never know their names, those who reviewed my book manuscript made many helpful comments and suggestions. I have tried to address the bulk of them, and I believe the book is much stronger for them. Likewise, the enthusiasm and ideas of Sandy Thatcher, then director at Penn State University Press, proved most valuable to me.

Some "behind the scenes" assistants deserve recognition on these pages. In Nicaragua, María Mercedes Rocha was my research assistant during much of the fieldwork in the countryside—and she put up with sleeping on that office floor and many other challenges in the process. I can only hope that she came away from the experience with more than bug bites and backaches. Tomás Morán accompanied me in the last month of interviews with the farmers, which included riding to outlying areas on mule back, something that I'm sure was more than he bargained for. And Carolina López labored over most of my tables, which I never would have had the patience to do. Back in the United States, Priscilla Bonnell "retouched" those tables with updated information in the final stages of preparing the manuscript. Jessica Rorem did a very professional job of preparing the maps of Nicaragua and Cuba. Tiffany Page assisted in multiple ways during several key stages of the preparation of the manuscript. Her calm, patience, and diligence were much appreciated. Two undergraduate research assistants helped out with some of the bibliographic research on Russia and China: Adam Fleisher and Carlos Almendárez, Jr. Thanks to all of you for your help on this project.

Several sources of funding were also crucial in making the research for this book possible. First, and most important, the Social Science Research Council provided me with a fellowship that made fieldwork in both the

Nicaraguan and Cuban countryside possible. Without it, the project would have never gotten off the ground. In later stages, the Center for Latin American Studies, the Committee on Research, and the Abigail Hodgen Publication Award, at the University of California, Berkeley, made funds available for further research (the first two) and the publication of my findings (the last). Given the difficulty of finding funding for research in Cuba, I am especially grateful to these institutions for making my fieldwork possible.

On a more personal note, my family and Edith Múñoz Díaz, who is a very dear friend, were all central to the completion of this project. My parents, Jean Gordon Enríquez and Eduardo Enríquez, have "kept the faith" throughout this long project that it would, indeed, eventually see the light of day—helping out with child care, moral inspiration, and general support. My brother, Robert Enríquez, generously brainstormed with me about the book cover. Edith played a crucial role in making it possible for me to carry out fieldwork under circumstances that were often difficult by taking care of my young son in the field with me during the early years of research for this book. Her support went so far as to accompany us to *donde el diablo perdío la chaqueta*—and to "protect" Adriano from *los armados* when she feared he might be at risk of being kidnapped. My compañero, Maurizio Leonelli, allowed me to take our baby . . . then toddler . . . and then young child on these research adventures, helping to equip our "team's" jeep with the many essentials for our travels in Nicaragua—potable water, food, and baby supplies—and to help out with our "move" from fieldwork in the Province of Havana to Santiago de Cuba, as well as offering many contacts, ideas, and much feedback, and constantly being confident that I would eventually dot the last "i" and cross the last "t" in its write-up. His undying optimism helped to keep me afloat at even the worst moments. Finally, our son, Adriano, who is now twelve years old, has "participated" in my various research endeavors and has endured quite a few experiences unimagined by many children in the Global North but are unfortunately all too common to children in the Global South. He not only survived them, but they have allowed him to become the worldly, compassionate, and astute observer of the social world that he is. Many, many thanks to all of you for contributing to this ultimately joint endeavor.

ABBREVIATIONS AND KEY TERMS

Acopio	state food crop purchasing and distribution agency (Cuba)
ALBA	Bolivarian Alternative for the Americas
BANADES	National Development Bank (Nicaragua)
CCP	Chinese Communist Party
CCS	Credit and Service Cooperative (Nicaragua and Cuba)
COMECON	Council of Mutual Economic Assistance
CPA	Agricultural Production Cooperative (Cuba)
ENABAS	National Enterprise for Basic Foodstuffs (Nicaragua)
FSLN	Sandinista National Liberation Front (Nicaragua)
granjas	state farms (Cuba)
ISI	Import Substitution Industrialization
kolkhozes	collectives (USSR)
manzana	1 *manzana* (mz.) = 0.7 hectares (Nicaragua)
Mercados Agropecuarios	farmers' markets (Cuba)
NGO	nongovernmental organization
NTAE	nontraditional agro-export
NTE	nontraditional export
parceleros	those who received a small plot of land to work in usufruct post-1990 (Cuba)
SA	structural adjustment
sovhozes	state farms (USSR)
UBPC	Basic Unit of Cooperative Production (Cuba)
UNAG	National Farmers' and Ranchers' Union (Nicaragua)
WTO	World Trade Organization

Introduction

The year 1990 was a turning point in both Nicaragua and Cuba. That year witnessed the electoral loss of the Sandinista National Liberation Front (FSLN), which had sought to move Nicaragua in the direction of socialism following the overthrow of the Somoza regime in 1979. The election results forecast a major shift in the government's reigning political economy, one characterized by a rapid retreat from socialism toward a "social economy of the market."[1] True to its electoral platform, the government that assumed power in 1990, which was headed by Violeta Barrios de Chamorro, quickly moved policy making toward a full embrace of capitalism. In Cuba, it was not a change of political regime that triggered the modification of that country's socialist political economy. Rather, the economic crisis that was provoked by the disintegration of the Council of Mutual Economic Assistance (COMECON—the Soviet-led trade bloc) forced the reconfiguration of what had heretofore been Cuban socialism.

The contrasting political economic orientations adopted in the 1990s by these two formerly allied governments resulted in the pursuit of distinct economic strategies. As part of Nicaragua's rapid retreat from socialism, a comprehensive program of economic stabilization and structural adjustment (SA) was put in place. In Cuba, the economic crisis led to the implementation of SA-like measures in several sectors of the economy. Yet policy in

1. See further UNO (1989).

some other sectors of the economy, such as agriculture, was notably different from that typically associated with SA. Logically, their respective strategies had varying consequences for certain sectors of their populations. One of these sectors was that involved in agricultural production.

The Chamorro government in Nicaragua largely patterned its policies according to the model set forth by the proponents of orthodox SA programs. As a result, it followed the path taken by most other Latin American governments beginning in the 1980s. The assumption behind that model was that the principal cause of the economic troubles facing many countries in the Global South was the economic approach they had employed for the previous few decades. That approach, known as import substitution industrialization (ISI), had prioritized moving their economies away from complete reliance on the export of primary products to also producing their own industrial goods, even if that required some degree of protectionism in the early stages of the push to industrialize. The state had played a crucial role in this effort to diversify production.

Proponents of SA argued that the Global South's nascent industrial sectors were being maintained artificially by the state, which was creating an unsustainable economic situation. By interfering with the market, these states were keeping afloat nonviable industries and causing disequilibria throughout their economies. The answer was, then, to restructure (or "adjust" the structure of) these economies so that they would once again experience growth, which was best done by emphasizing what they had produced historically—products for export.

Within Nicaragua's SA program, export production was to be prioritized. Within agriculture, export products were to be the focus of attention. As a result, domestically oriented production—which was the specialty of small farmers, who made up close to 90 percent of the country's agriculturalists in 2001[2]—found itself relegated to a distant second place in terms of the government's resource allocations. This situation, combined with other aspects of SA, led to the increased marginalization of this sector of producers. Their marginalization translated into a growth in the numbers of rural dwellers living in poverty and to the concentration of land, the key indicator of wealth in the countryside.

Conversely, within the context of the reform that was designed to address Cuba's economic crisis, agricultural producers—particularly those who em-

2. CIPRES (2007, 96).

phasized food production—became a select group of beneficiaries of the government's limited resources. This was, in good measure, a consequence of Cuba's need to achieve greater self-sufficiency, especially in food, given the extreme shortages that resulted when its trade relations fell into disarray. Moreover, through various means, much of Cuban agriculture was downsized (i.e., the scale of units of production was reduced) and smaller-scale production gained new prominence in policy making toward this sector. Hence, the cooperative and small private farm sectors of agriculture—which were responsible for approximately 80 percent of the production of Cuba's principal food products in 2005 (calculated from Tables 5.4.1 and 5.4.2)—were given priority. Their relatively better-off economic situation compared with most of the rest of the Cuban population was the logical outcome.

Thus, in the 1990s, these two countries' governments adopted very distinct approaches toward their respective rural sectors. Earlier on, their socialist-oriented governments had implemented far-reaching agrarian reform programs, involving land redistribution, the formation of state farms, and the provision of loans to small producers to redress the previously existing inequities in their agrarian sectors and to bring about more balanced development. However, with the shifts in overall orientation reflected in each government's economic policies in the 1990s, the status of the small farm sector was once again in question.

The status of small farmers has relevance beyond the issue of the social and economic well-being of this one sector of the population, with ramifications throughout their societies and even extending into others. Since, in many countries in the Global South, they are central actors in food production, their labors can contribute in crucial ways to efforts to move a country toward greater food self-sufficiency and, therefore, greater independence. The significance of this only became clearer in recent years as food prices rose in the international market because of the growing emphasis on biofuels and climatological effects on production, among other factors. In addition, in many countries, small farmers continue to represent a sizable part of the labor force, which means that their economic viability as producers affects a notable percentage of the population. Unless other sectors of the economy can absorb their labor—which is not typically the case in the Global South—if they are pushed out of this activity, they come to find themselves among the multitudes of unemployed and underemployed urban slum dwellers and migrants in search of employment in the Global North. Thus, the implications of their ability, or lack of it, to subsist as farmers in the Global South are very broad in scope.

This book will examine the nature of the Nicaraguan and Cuban governments' differing political economic transitions, the effect policy shifts have had on their small farmers, and the initiatives set forth by these same farmers to ensure their own ongoing production and survival. It will also analyze each government's vision of development and the role of the small farmer within that vision.

Interestingly, Nicaragua's and Cuba's approaches to agricultural producers, and especially their small farmers, bear some important resemblances to those of Russia and China, respectively.[3] In Russia, agricultural policy changes have been key in the full-scale retreat from socialism since 1990. Those changes have had, by and large, an extremely negative effect on the sector, leaving it to slide toward mere subsistence production. And those producers associated with the former state and collective farms, who generate more than half of the country's food products on their small parcels of land, have borne the brunt of this effect. They have seen their lives and livelihoods become more marginalized. In China, in contrast, major changes in the agricultural sector have been of a piece with the larger endeavor of reconfiguring socialism. In the initial phase of this latter country's process of change (1978–85), the rejuvenation of agriculture was seen as central to China's economic reactivation. The entire farming sector was reorganized into individual small-scale units that were responsible for producing all of the country's food products. And, farmers were the beneficiaries of this wholesale policy shift, as their rising income demonstrated.

To locate the Nicaraguan and Cuban cases within a larger context of transition, they will be compared with Russia and China. I use the position of the small farmer as an index of these states' different approaches to the expansion of market relations in their formerly socialist economies. (See Appendix A.) As I will show, this sector has been hard hit where a rapid retreat from socialism was embarked on—as in Nicaragua and Russia. At the same time, it has flourished, relatively speaking, where socialism was reconfigured—as in Cuba and China. Clearly, from the point of view of small farmers, continuing some form of socialism was more advantageous than a complete embrace of capitalism.

3. See Appendix A for a discussion of the applicability of the term *small farmer* in each of these contexts.

This conclusion—the evidence for which will be presented in the following chapters—finds a strong echo in the theses set forth by Karl Polanyi (1944) in his study of the social and economic effects of an earlier worldwide expansion of market relations, that is, the original period of liberalism. That study argued that the unfettered expansion of market relations in England was highly destructive for rural small-holders, effectively eliminating their livelihoods and transforming them into the nascent industrial working class. The conditions they found themselves in once this transformation was finished were similarly troubling, to the extent that Polanyi (1944), following William Blake, referred to their new places of livelihood as "the Satanic mills."

The insights gleaned from my comparative study will be set against the backdrop of Polanyi's (1944) analysis of the spread of market relations. His study, however, speaks to a number of issues that a few other bodies of literature also raise. One of these has emerged over the past couple of decades and is focused on the nature and consequences of the changes experienced in China, Russia, and a number of other countries as they have moved away from what had been orthodox socialism. Several questions have been at the center of this debate,[4] only one of which is particularly relevant here: what is the relationship between the expansion of markets and inequality? As capitalist relations spread into these formerly redistributive economies, has inequality increased or diminished?

In addressing these questions, I will also draw on the more general literature on the social impact of the expansion of market relations inherent in the recent neoliberal era. Although this literature is more inclusive than that describing the transition from socialism, it has insights to offer for the study of the latter.

This study looks at what has happened in the agricultural sector as this process moved forward. More specifically, it assesses the effects of the transition from orthodox socialism on small farmers. As such, it also engages with the long-running debate on the future of the peasantry or small-scale producers. In the late 1800s, F. Engels (1977) and V. I. Lenin (1957) posited, albeit in different ways, that the expansion of capitalism into agricultural production would lead to the disappearance of the peasantry. Yet some of their contemporaries, and others in more recent decades, have disputed this position. Instead, they have suggested that the peasantry—at times with the

---

4. Szelenyi and Kostello (1996) describe these questions.

assistance of other actors, such as the state—will continue to maintain a foothold in agriculture, even as its access to productive resources is reduced.

By looking in-depth at the circumstances of small farmers in two countries that have moved away from orthodox socialism, but in distinct directions, my study is located at the crossroads of these various literatures. The comparison with Russia and China not only highlights the commonalities of these two pairs of cases[5] but also lends weight to my arguments regarding the question arising at this theoretical intersection.

What follows below draws on findings generated by research I have conducted on these four countries (see also Appendix B for a discussion of my methodology). The case studies of Nicaragua and Cuba are based on interviews carried out with representatives of the key actors in each country's agricultural sector—the respective agricultural ministries, farmers' unions, nongovernmental organizations with projects in rural areas, and research institutes engaged in the study of related issues. These informants provided general characterizations of changes in agricultural policy and the implications these would have for the rural population, especially small farmers. These were complemented by the implementation of a survey of small farmers in four rural municipalities in each country.[6] In Nicaragua, these producers were farming on an individual basis at the time of the interview, although some of them had previously been members of the Sandinista Agricultural Cooperatives (CASs) that had formed in the 1980s on agrarian reform land, while others continued to be members of Credit and Service Cooperatives (CCSs).[7] In Cuba, the survey drew on farmers participating in CCSs, Agricultural Production Cooperatives (CPAs), Basic Units of Cooperative Production (UBPCs), and the *parcelero* sector (which included those who had

5. There were major differences between these regimes (of both a political and economic nature), and at least one of them—Nicaragua—was not explicitly socialist. In the latter case, the stated goal for the transformation that was initiated there in 1979 varied from more socialist to more social democratic in nature. Without delving into a nuanced discussion of all of the elements that compose a "socialist" regime, their ostensive prioritization of workers and peasants, their generally redistributive orientation, and their willingness to intervene significantly in capitalist production relations when these undercut the former priorities will suffice for now to justify the label their leaders used regularly (in the cases of the USSR, China, and Cuba) or occasionally (in the case of Nicaragua) to describe their states.

6. The definition used for *small farmer* in each case needs clarification. Traditional definitions of producer size were employed to select these farmers, which were based on landholding size. Yet the precise definition for this category varied by region in each country, taking into account access to infrastructure, markets, and so on. These specifications will be detailed when each region studied is presented in Chapters 4 and 6.

7. Only a very small minority of Nicaragua's farmers worked their land collectively by the late 1990s.

received a tiny plot of land to farm as individuals during the 1990s).[8] In addition, in Cuba, I interviewed one or more officials from each of the various CCSs, CPAs, and UBPCs whose members participated in the survey.

Once the fieldwork in Nicaragua and Cuba was finished, I engaged in a comparison of these two cases with those of Russia and China. Thus, I undertook a comprehensive review of the secondary literature on the changes that have occurred in each of these latter two countries in the wake of their own reform processes.

I will begin this examination of different approaches to market expansion by exploring, in Chapter 1, how the issues they raise are situated within the relevant theoretical literatures. This theoretical discussion will be followed, in Chapter 2, by a comparative analysis of Russia and China. Hence, as I embark on an examination of Nicaragua and Cuba in the subsequent four chapters, both theory and larger cross-country comparisons will serve as the background. Upon completing this exploration of the effect that post-1990 political economic changes have had on small farmers in Nicaragua and Cuba, in the Conclusion, I will analyze the similarities and differences they share with the cases of Russia and China, respectively. This will help to highlight the lessons to be derived from comparing these distinct approaches to the market.

8. In Cuba, I also included both individual farmers and cooperative members in this "small farmer" category. In Chapter 6 (n. 1) and in Appendix A, I explain the logic for this.

PART 1

Transitions from Socialism and Their Social Consequences

1

The Theoretical Backdrop

Profound economic crises characterized Nicaragua and Cuba in the early 1990s. Despite their sharing a socialist orientation toward policy making during the 1980s, the strategies that each government later pursued to address the crisis differed significantly. These strategies reflected the overall political economic orientation of their respective governments.

In Nicaragua, the structural adjustment implemented in response to the crisis was geared to open up the economy to the international market and to eliminate all barriers to the market that might exist. A central part of this effort was the renewed emphasis on production for export. Traditional and nontraditional agro-export products were to be prioritized, as well as nonagricultural exports. Domestically oriented production, and producers, had to quickly find means of vying against the massive influx of imported goods or perish. Given that most small farmers engaged in domestically oriented production, they were subjected to the fierce winds of international competition at the same time as they were cut off from receipt of any of the government's increasingly limited forms of support for agricultural production. The result was that they (as producers) and their production became more marginalized. Many, if not most, of them were forced to withdraw into subsistence farming. Nicaragua's new policy makers did not look to this sector of production to play any kind of major role in agricultural development or in economic development more generally, and their policies toward small producers reflected this.

In contrast, with the economic crisis facing Cuba's policy makers in the early 1990s came the recognition that bold initiatives would have to be adopted if the regime, and Cuban socialism with it, were to survive. An early manifestation of the crisis was the shortfall in food imports, as well as of the imported inputs required for Cuba's large-scale agriculture. Hence, local food production became a priority, especially over the next few years, as did those who were most efficient at its production—small farmers. Many incentives were set in place—including the most critical of all, being given access to land—for those who were willing to move into agriculture from other sectors of the economy, triggering peasantization.[1] Far from considering small farmers to be only social subjects whose welfare would have to be protected to some extent, in the 1990s, Cuba's policy makers also came to consider them to be important economic actors with the potential of contributing to the country's agricultural—and, therefore, economic—development.

The transformations Nicaragua and Cuba underwent in the 1990s paralleled changes set in motion in two significantly larger socialist states on the other side of the globe, Russia and China, respectively. Russia engaged in a rapid retreat from socialism with the collapse of socialism there and in Eastern Europe. Structural adjustment (SA) measures were adopted in all sectors of the economy, with agriculture being among the most hard hit by them. As a consequence, output plummeted, and a return to subsistence production was widespread. Clearly, only those who were willing to take up the challenge, in risky and difficult times, to become capitalist farmers were perceived to be economic actors. The remainder were left to fend for themselves.

China's policy makers saw their peasantry in quite a different light. Given their sheer numbers, and their centrality in ensuring political and economic stability, their role in reconfiguring socialism, which was initiated shortly after the death of Mao Zedong, was taken much more seriously by China's reformers. Although agriculture was not envisioned as a leading sector of China's new economy, its crucial contribution to that economy was not underestimated. Thus, a variety of means were adopted to try to improve the performance of this sector and with it the lives of the multitudes of China's small farmers. These began with increases in the prices paid for agricultural produce and improvements in the terms of trade between agriculture and

---

1. The term *peasantization* is used herein to refer to the movement of people who had not previously taken part in agricultural production into it.

industry and went on to include the momentous downsizing of agricultural production to the household level. That China's policy makers considered their small farmers to be economic, as well as social, subjects was evident in this whole package of policies.

Chapter 2 seeks to place Nicaragua's and Cuba's small farmers in a comparative light, as their governments pursued differing pathways toward the market. Hence, in preparing to look more closely at the Nicaraguan and Cuban cases, which will follow in subsequent chapters, I will discuss the varying circumstances of rural small-holders in Russia and China as each embarked on opening up to the market.

However, the transition toward the market that all of these countries have been experiencing can also be located within a theoretical context. The overarching discussion that this book is engaged with concerns the social and economic effects of expanding market, or capitalist, relations. This discussion contains within it debates about "the transition to the market" taking place in former socialist countries, as well as the now historic debate about the future of the peasantry: will the expansion of market relations, or capitalism, inevitably bring about its disappearance? Reflection on the insights derived from my in-depth examination of Nicaragua and Cuba, as well as their comparison with Russia and China, will be framed by these several areas of debate to address the effect on small farmers of these distinct transitions to the market. This chapter will sketch out these debates and how they intersect in countries transitioning toward the market.

The Relevant Theoretical Debates

Analysis of the expansion of market relations and its social and economic effects began long before the recent period of "transition" experienced by countries formerly characterized as socialist. Among those to study the way in which this unfolded much earlier on was Karl Polanyi (1944). He focused his analysis on England and the process by which market relations, or capitalism, took hold there in the nineteenth century. That is, he was concerned with the first era of liberalism.

During that era, relations of production in the countryside were transformed as small-holders, who had previously sustained themselves through subsistence production, grazing their livestock on "the commons," and obtaining cooking fuel in "the forests," were shut out from these various sources of livelihood through the Enclosure Movement. The Enclosure Movement represented a major step in the direction of defining and delimit-ing—especially in the sense of erecting barriers around—the "private prop-erty" of England's landed nobility. Although evolving in fits and starts, the movement eventually succeeded in cutting peasants off from any means of production, thereby forcing them to leave the countryside and search for new livelihoods elsewhere. That "elsewhere" most commonly ended up being the nascent factory towns at the forefront of the Industrial Revolution. Thus, they came to form the workforce that was needed to make that revolution possible.

Polanyi (1944) stressed that this was not a "natural" process but rather a process in which the state played a fundamental role in opening the way for it. In addition, he highlighted the growing immiseration of the poorer sec-tors of society as they were stripped of their access to means of production and subjected to the "Satanic mills" that the Industrial Revolution gave rise to. The resulting immiseration was both material, in terms of the intense poverty and degradation of living conditions that came to characterize their lives, and cultural, in the sense of having their entire world torn asunder.

While these changes were fundamentally altering the shape of the En-glish countryside, the Industrial Revolution and the liberal economic era, of which it was the centerpiece, were likewise altering social and economic dynamics outside of England. As Polanyi (1944) argues, international free trade, the hallmark of liberalism (and of neoliberalism, later on), represented a serious threat to agriculture and its producers. In countries where the state was relatively strong, such as in Central Europe, laws were enacted (e.g., the Corn Laws) to protect agriculture and those who engaged in it from the effects of cheap grain imports. But "the politically unorganized colonial peo-ples" were not able to protect themselves in a similar fashion (Polanyi 1944, 183). Hence, the blows wrought by international free trade were felt much more acutely in the colonies.

Yet Polanyi (1944) also emphasized that the spread of market relations—what he terms "movement"—always and everywhere provoked a "counter-movement" in the form of protective measures taken by society. At times, these countermovements were embodied in actual social movements and at other times, in actions taken by the state (e.g., in the form of new laws). Regardless of the shape they took, this "double movement" underlay the "dynamics of modern society" (Polanyi 1944, 130).

A number of the issues Polanyi (1944) raised with regard to the first era of market expansion are themes central to the debates about the more recent period of transition to the market, especially, but not only, that experienced where orthodox socialism had previously prevailed. These latter debates were initiated in response to the transformation China's political economic model has undergone since 1978. They have been concerned with both the more general and the more specific consequences of that transformation. Thus, analysts have sought to define whether China is in transition to capitalism, with some arguing that there is essentially one path to get "there," with the social and political characteristics of "there" being uniform across societies (e.g., Nee 1989, 1996); others believe that China, like other former socialist republics, will develop its own model of capitalism, with a plurality of such models emerging (Stark 1996). Still others suggest that perhaps depicting these countries as "socialist" or "capitalist" should not be a foregone conclusion and that this depiction should be subject to examination.

Rather than getting immersed in this large question, I will take for granted that these countries are not all following the same course toward the market. Some of them have embraced a capitalist political economic orientation wholeheartedly, seeking to leave no element of their former redistributive model intact. In contrast, others still claim to be socialist in orientation, while having permitted the market to enter into certain parts of the economy and society. Szelenyi and Kostello (1996, 1087–94) argue that market (or capitalist) penetration can take place in the form of (1) local markets in otherwise redistributively integrated economies, (2) socialist mixed economies, or (3) capitalist-oriented economies. They locate post-1990 Russia and Eastern Europe in the last category, while China between 1978 and 1984 fit in the first and then shifted to the second in 1985. I find this conception compelling and would only suggest that it might be useful to think of the first and second of these categories as representing these states' efforts to "reconfigure" socialism, while the latter represents a full-scale retreat from it. In expanding this discussion to include Cuba and Nicaragua, I would situate Cuba in the area of reconfiguring socialism and Nicaragua in the area of retreating from it (see Figure 1).

But, it is important to include an addendum here. As Polanyi (1944, 207) noted, not all countries are positioned equally within the international economy.[2] Writing of the liberal era, Polanyi stated that "the world counted a

---

2. The academic field of development emerged, in fact, in the aftermath of World War II to seek an explanation for this situation and to set forth proposals for its modification.

Fig. 1 Pathways to the market

	Rapid retreat from socialism[a]	Reconfiguring socialism[a]	
	Capitalist-oriented economies[b]	Local markets within redistributive economies[b]	Socialist mixed economies[b]
Practically self-sufficient[c]	Russia	China (1978–1985)	China (post-1985)
Exporting countries[c]	Nicaragua	Cuba	

[a]Enríquez.
[b]Szelenyi and Kostello 1996.
[c]Polanyi 1944.

limited number of countries, divided into . . . exporting countries and practically self-sufficient ones, countries with varied exports and such as depended for their imports and foreign borrowing on the sale of a single commodity." This characterization is still relevant today, although perhaps not as extreme. Thus, I would describe Russia in the contemporary period as a "practically self-sufficient" country engaged in a rapid retreat from socialism, the latter category of which it shares with Nicaragua, an "exporting country." China has been a "practically self-sufficient country" engaged in reconfiguring socialism, while Cuba, pursuing the same political economic pathway as China, is an "exporting country." Hence, at the same time as I seek to highlight these paired countries' similarities, I do not wish to deny their differences.

Moving beyond how exactly these societies should be thought of, this literature is also concerned with the societal consequences of these transformations. Victor Nee (1989) was among the first to initiate this debate. He set forth an argument that drew heavily on modernization theory in depicting the transition under way in China as one "to capitalism." Moreover, China's transition had a number of features in common with the "original" transition from traditional to modern society. As suggested earlier, Nee (1989) saw China as on the path to capitalism, and it was a path that would lead to a reduction in inequality. Over time, this last issue, the transition's effect on societal inequality, has become a centerpiece of larger debates about this process of change.

Nee's (1989, 1996) exploration of this issue was focused on the economic position of Communist Party cadres and how it was changing as the process of transition moved forward. He was especially interested in their standing in regard to the new "entrepreneurial" sector. He posited that as the transition progressed, the economic advantages of the cadres would diminish and the earnings of the entrepreneurs would increase, leading to less income inequality overall.

This thesis, however, has been strongly contested by both China experts and those whose work examines other regions. Xie and Hannum (1996, 950) use a distinct quantitative data set and a different approach to analyzing it. They conclude that "overall earnings inequality remains low and only slightly correlated with economic growth" (growth is the proxy used for economic "modernization" or the expansion of market relations). Furthermore, economic growth did not alter the earnings differentials between those who were party cadres and those who were not. Szelenyi and Kostello (1996, 1088) hold a contrasting position. They argue that, during the period when local markets are permitted to function within a larger redistributive economy, inequality diminishes somewhat as peasants see their incomes rise. But during the period of socialist mixed economy, when markets have expanded into ever-greater areas of the economy, such as in post-1985 China, there are large increases in inequality. These large increases in inequality have also characterized the capitalist-oriented economies of Eastern Europe since 1989.

In a comparison of the transformation of China's and Russia's political economies, Nolan (1996/1997) notes that new inequalities resulted from the growth of market forces in China, during what Szelenyi and Kostello (1996) call its socialist mixed economy phase. Yet in analyzing Russia's full embrace with capitalism, he describes the significant redistribution of income and wealth that took place, leading to a dramatic increase in inequality. Citing data from the journal *Trud*, Nolan (1996/1997, 236) specifies the magnitude of this increase in Russia: "The ratio of the income of the top decile to that of the bottom decile was in the order of 1:5.4 in 1991, and had [already] risen to 1:8.0 by the end of 1992." This trend continued through the rest of the decade and well into the next one, as can be seen in the following observation by Wegren, Patsiorkovski, and O'Brien (2006, 372–73), quoting a Russian source: "the Gini coefficient, which is a statistical measure of inequality, increased from 0.28 in 1992 to 0.38 in 1995, and increasing to 0.41 ten years later." That is, where a capitalist orientation was adopted by the state, inequality grew by leaps and bounds.

Shifting the focus to some smaller states in which markets have expanded where they were previously almost nonexistent or somewhat restricted, Latin America offers some interesting cases for examination. Cuba now logically fits in Szelenyi and Kostello's (1996) category of socialist mixed economies, what I refer to as a country in the midst of reconfiguring socialism. This is a situation in which market and redistributive systems coexist under the hegemony of the latter. True to what their model would suggest, inequality has increased substantially since 1990, when Cuban socialism began to be transformed (see Quintana Mendoza 1996; Espina Prieto n.d.; Dilla 1999; Ferriol Muruaga 1998). Among the principal factors, if not the principal factor, contributing to this phenomenon was the legalization of holdings in U.S. dollars by the Cuban population (which occurred in 1993), combined with the more general "dollarization" of the economy. This put at tremendous advantage those who had any income in U.S. dollars.[3]

However, as one would expect in the case of a socialist mixed economy, there were still many social policies in place that served as at least a partial counterweight to the tendency toward social differentiation. These included the strong commitment to maintain public systems of education and health care that were comprehensive and free for all Cubans. In addition, a food rationing system continued to be in place, although it no longer covered all of the needs of the country's consumers. These various social guarantees provided Cuba with among the highest social indicators in the region (PNUD 2002; CEPAL 2000), as well as with a special comparative advantage in terms of its level of human capital.

Even though certainly not alone in distinguishing Cuba from many other countries in the process of allowing the entry of market relations into sectors of the economy formerly controlled by the state, these social guarantees were important in keeping the levels of social inequality lower than they might otherwise have been. At the same time, the distinct nature of Cuba's economic reform—that is, it not having pursued the orthodox model of SA—was also crucial in producing this outcome.

In contrast, most other countries in Latin America—including former socialist ones and others that, while not socialist in nature, had used various

---

3. In November 2004, the U.S. dollar was banned from all commercial transactions (*New York Times* 2004; Thompson 2004). After that, Cubans were "encouraged" by a 10 percent exchange fee on the U.S. dollar to use Cuban pesos, convertible pesos, and Euros. The measure was designed to protect the economy from tightened U.S. sanctions and to build up its hard currency base (Spadoni 2004). Nonetheless, having some income in a foreign currency still enabled the purchase of many goods that were otherwise unavailable. Even with this new restriction, foreign currency recipients continued to be at an advantage.

policies that protected key sectors of the economy from the full effect of interchange with the international market—have opened their doors to the advance of market forces through adoption of orthodox SA programs. Chile and Nicaragua serve as two examples of countries whose governments had, for a time, sought to approach policy making from a socialist mixed economy orientation. Yet with the subsequent change in regime in each case, the new team of policy makers initiated a rapid retreat from socialism, employing, among other measures, a standard structural adjustment of the economy to ensure that capitalism was fully embraced. As predicted by Szelenyi and Kostello (1996), the result was increased inequality.[4] In describing the situation in Chile, Altimir (1994, 95) says that "the inequality grew at a rate that was probably without precedent." Nicaragua also experienced a dramatic increase in inequality, with available statistics only confirming what was evident to even the most casual observer.

On the basis of the evidence describing China, Russia, Cuba, Chile, and Nicaragua, but especially the latter three countries, I would concur with Szelenyi and Kostello's (1996) position concerning the socioeconomic effect of opening up to the market. In the two cases in which this process went the furthest—Chile and Nicaragua—and a capitalist orientation was fully adopted, inequality increased exponentially. In Cuba, where the process was more graduated and some important redistributive mechanisms were still in place, increases in inequality were much less extreme. Clearly, whether the hegemonic orientation remained redistributive was a key determinant in the level of inequality produced by the economic policies implemented in these countries. Where it continued to be predominant, the effect of economic reform was not as severe for these societies' less advantaged than where it had been replaced by a capitalist orientation.

Elsewhere in Latin America structuralist policies promoted by the UN's Economic Commission on Latin America (CEPAL) had been common in the several decades preceding 1980, even though socialist economies had never gained the day. This meant that some sectors (if not all) of domestic industry were protected from the international economy by high tariffs on imported goods, and domestically oriented production was facilitated with economic incentives (i.e., the import substitution industrialization strategy mentioned in the Introduction). However, the debt crisis that emerged in

4. For references on this process in Chile, see Scott (1996) and Altimir (1994); and in Nicaragua, see Arana and Rocha (1998) and others to be discussed in Chapters 3 and 4.

the late 1970s opened the way for international lending institutions to insist on bringing an end to such protective measures and on placing a new emphasis on interchange with the world market according to these countries' special comparative advantages. The social consequences of these macroeconomic shifts typically included an increase in poverty and growing income inequality.[5] Whereas earlier Chile and Nicaragua had engaged in an expansion of social services along with other redistributive efforts, after their period of socialism, they joined with the rest of those Latin American regimes structurally adjusting their economies in significantly cutting back on state-provided health care, education, and pension plans. Thus, as times got harder for most of these countries' populations, the social safety net that had previously been available to them also shrunk, if not disappeared.

With growing inequality resulting from an expansion of the market as the general context, this study is particularly concerned with the effect this has had on agricultural production and those who undertake it. One sector of producers—small farmers—is so situated that closely examining their situation provides a unique window into the dynamics of distribution where markets are spreading in the contemporary era. In part, this is the case because this sector has not, typically, been the most important sector for export production. Hence, the productive and living conditions of small farmers in the midst of market expansion speak clearly to the issue of winners and losers in this economic environment.

Moreover, the well-being of small farmers in the Global South has broader significance both within their own countries and beyond them. Aside from their contribution to domestic food production, which is often quite substantial, this sector frequently absorbs a notable part of the labor force. Where other productive sectors are not in a position to incorporate them should farming cease to be economically viable, the resulting discontent, potentially producing unrest in the countryside, may come to represent a political problem for the regime in power. The emergence of Vía Campesina—an organization of small farmers from fifty-six countries in Latin America and the Caribbean, North America, Europe, Africa, and Asia[6] that formed in 1993 in response to the dislocations experienced by those within this sector—provides an example of this dynamic at work. They may also

---

5. This pattern is documented in various essays included in Bulmer-Thomas (1996). See especially those by Fitzgerald, Panuco-Laguette and Szekley, and Ferreira and Litchfield. See also Korzeniewicz and Smith (2000) and Huber and Solt (2004).

6. See further, Borras (2008, 26) with regard to Vía Campesina.

come to play a role in urban unrest, as they join the swelling ranks of the urban poor. And, finally, if neither the countryside nor the city can provide them with a livelihood, international migration within the Global South or to the Global North becomes the next logical option for them. In sum, examination of this sector can shed light on multiple additional social issues.

This assessment of the situation of small farmers in the neoliberal era will be brought to bear on a debate stemming from a much older literature, analyzing the effect on the peasantry of the expansion of capitalism into agriculture. Dating back to the work of Engels (1977), Kautsky (1976), and Lenin (1957) in the late 1800s, the debate it has given rise to has continued to the present time. Its essence concerns whether the peasantry are doomed to extinction as a consequence of capitalism spreading into agricultural production. Lenin (1957) argued that, as agriculture became increasingly commercial, the middle peasantry would experience differentiation, with a small part of its members becoming rich peasants—and eventually capitalist farmers (or a rural bourgeoisie)—and most becoming poor peasants—and eventually landless agricultural workers. Subsistence production, and even petty commodity production, would give way when confronted with capitalist agriculture, leaving in its wake two classes in the countryside: the rural bourgeoisie and the rural proletariat.

Although Polanyi (1944) did not intend to engage with Lenin, he also weighed in on this issue. Through his analysis of the process by which a working class was formed, he reached a conclusion about the destructive capacity of market relations that was similar to Lenin's. Effectively, the successful transformation of labor and land into commodities (albeit "fictitious" ones), which was essential for the spread of market relations, required the "release" of the poorer rural strata from their ties to social and economic institutions in the countryside. For Polanyi (1944), the elimination of this social category was both a precondition for, and a result of, the expansion of market relations.

The "agrarian question," as this debate has come to be termed, has been taken up again more recently in the context of both more as well as less developed countries.[7] In a more nuanced, but similar vein to Lenin, de Janvry

---

7. This "question" has also been the topic of largely theoretical discussion. Akram-Lodhi and Kay (2008) bring together a variety of perspectives on the relevance of this issue in the context of twenty-first-century globalization. These perspectives include, on the one hand, scholars who argue that the question needs to be reframed to account for the global nature of capital and the reduced significance of the nation-state in this new era, as well as the international response of the peasantry who are calling for a new kind of agriculture (McMichael 2008a), to be discussed later. On the other hand, this theoretical discussion also includes those who posit that the classi-

(1981) argues that the peasantry in Latin America has come to have an important function within agricultural capitalism: it is the key provider of cheap food for urban workers and cheap labor for large-scale capitalist farms. That is, the peasantry engages in production for the market, but it is forced to sell its principal product, food crops, at very low prices. Low prices, in combination with some other factors, lead to the social differentiation foreseen by Lenin. Thus, the peasantry's access to agricultural resources becomes increasingly limited, which inhibits it from surviving completely on its own production. Increasingly, it is forced to sell its labor power on the agro-export estates part of each year. According to de Janvry (1981), this relationship, which he calls *functional dualism*, allows for the expansion of capitalism into Latin American agriculture. Yet, as it expands, it simultaneously undermines the basis for functional dualism through the growing marginalization of the peasantry. In sum, the peasantry will become so marginalized it will cease to play even this restricted role in agricultural production.

However, other studies have shown that this is not necessarily an inevitable outcome, nor a linear process, either in the Global South or in the Global North. Examining the situation of wheat farmers in nineteenth-century United States, Friedmann (1978) demonstrates that large-scale capitalist wheat farms actually gave way to household production for the national and international market. Several factors were at play in creating this situation. With the emergence of a world market for wheat in the 1870s, the international price for this commodity fell. In the same general period, mechanization of wheat planting and harvesting occurred, which made it possible for large tracts of land to be farmed by just a few people (i.e., by household labor). Given that household production does not require the generation of a profit above and beyond what those who perform the labor require for their own reproduction, in contrast to capitalist production, this provided it with an economic advantage. In addition, although capitalist production does have means for constraining the consumption of workers, especially if there is an abundance of available laborers, household production is even more capable of doing so. Moreover, the existence of an "agrarian frontier" in the West gave the landless the option to migrate to where they could farm their own tracts of land instead of submitting themselves to employers' ef-

---

cally posed question is still valid, albeit with differences as to where the emphasis should be placed—on labor (Berstein 2008) or on capital (Akram-Lodhi, Kay, and Borras 2008). Regardless of their differences—some of which are quite substantial—the publication of this volume illuminates continuing scholarly concern with the "agrarian question."

forts to lower their wages. As a consequence, wages rose for those who were still willing to sell their labor. This put capitalist production at a further disadvantage.

Aside from these more technical and social factors, it is important to highlight the role of the state in this phenomenon. As Friedmann (1978, 582) notes, "the availability of land was not a 'natural' fact." It was a result of the state's expansionist policies, leading to the forced removal of the indigenous population that had previously dwelled there. The state promoted the westward movement of settlers, as opposed to absentee capitalist farmers, who would ensure the territorial integrity of the newly conquered lands. Likewise, the settlers were offered bank loans to facilitate their farming efforts with the goal of having their production shipped on the heretofore underused transcontinental rail lines.

According to Friedmann (1978, 583–84), similar agrarian frontier/settler dynamics existed in Canada, Argentina, and Australia, leading to an increase in world wheat acreage by almost half by 1934. In contrast, she shows that among the more traditional wheat-producing countries of Europe, it was only where the state set in place protective tariffs that capitalist wheat production was able to survive. Hence, technological innovations, social relations, and state intervention had resulted in household wheat production supplanting capitalist wheat production in most parts of the world by 1935.

Issues of technology, social relations, and the state also came to play a role in the uneven spread of capitalism in the production of another crop in the nineteenth-century United States: cotton (Mann 1990). While capitalist production relations came to predominate in cotton production in the Southwest, in the Old South, noncapitalist relations prevailed until the midtwentieth century. The devastated condition of the Southern economy in the wake of the Civil War meant that capital for production was scarce. Among other things, this complicated the possibility of employing wage labor. Moreover, given the emancipation of the slaves, without a great investment in management, it proved difficult to recruit and control workers. At the same time, landowners were reluctant to assume all of the economic risk that production implied in the context of their own weakened economic position. Finally, even if producers had been able to overcome the financial obstacles to purchasing the machinery employed in Southwest cotton production, the topography of the Old South—its rolling hills contrasted sharply with the flat plains of Texas, Arizona, and California—and the region's nonuniformly sized fields made use of it impossible, thereby eliminating this solution to the "labor" problem. All of these factors led producers to rent out their land to sharecroppers and tenants. Thus, production relations for the same crop varied by region for the better part of a century.

With the increased interventionism of the state during the Great Depression, the circumstances that had led to this variation began to change. To save the country's farmers from bankruptcy, the national government began to pay subsidies to them for the production of certain crops and for restricting production of others. Cotton was among the crops affected by these subsidies. However, it was landowners, not sharecroppers or tenants, who received the subsidies. As a result, landowners found themselves with an infusion of financing, which made them less dependent on low-cost forms of farming their land. In addition, the introduction of synthetic fibers and international market pressures encouraged them to look to alternative commodities to replace cotton production entirely.

Mann (1990) set forth her account of production relations during this period to illustrate her argument that the expansion of capitalism into agriculture is an uneven process and that the elimination of noncapitalist production relations is not an automatic phenomenon.[8] This case clearly put in evidence that cotton production had not yet been captured by capitalism in the Old South, even though it had been in the Southwest.

In analyzing "the agrarian question," various studies have approached it from the capitalist landowner's point of view (or capital, writ large) and what made sense for him or her. But research has also been done that addresses this question from the point of view of the peasant, or small farmer, whose continued existence is debated. For example, Deere (1990) examined the transition to capitalism in northern Peru. She found that, as peasants' access to agricultural resources became increasingly constrained, they used a growing number of strategies among their household members to "reproduce the peasant household as a unit of production and reproduction" (Deere 1990, 2).[9] Whereas some household members continued to work on their tiny,

8. In a study of strawberry production in California, Miriam Wells (1996) likewise found that household production in the form of sharecropping had gained the day during the mid-twentieth century. For a variety of social, economic, and political reasons, it made sense not to use capitalist relations of production.

9. Kay (2006) notes that the recently coined term *new rurality* refers to the multiactivity of peasant farm households in the neoliberal era, as they struggle to survive in this increasingly hostile economic environment. Yet Deere's (1990) account highlights that this is not a novel strategy of survival.

mostly borrowed, plots of land, others went off to the coast to work as wage laborers who were then able to contribute periodically to the household income, and others engaged in petty commerce, traveling over great distances to purchase and sell their wares. Together, they were able to keep hearth and home intact.

Although Deere underlined the fact that through such means the elimination of the peasantry by capitalist relations of production was greatly slowed, she did not end her argument there. Rather, she posited that it was essential to look inside households to identify the multiple class relations they contained. In all likelihood, some household members participated in what she referred to as "fundamental class processes," which involved those who, through their own production, generated surplus value and those who extracted it from the direct producers. At the same time, other household members engaged in subsumed class processes, which involved people who partook in distributing surplus value (such as merchants, moneylenders, and landlords). Moreover, individuals could occupy multiple class positions based on their participation in these distinct classes' processes. Deere (1990) believed that it was necessary to go beyond the statement that household production continued to exist in northern Peru despite the advance of capitalism. In addition, the different economic activities of its members had to be analyzed. Hence, it became possible to appreciate that the operation of distinct class processes within the peasant household went a long way to explaining its survival.

Finally, in a different tack on the same issue, Anthony Bebbington (1998, 1999) and others pointed to another means that some small farmers used to assist them in surviving in adverse circumstances. In examining the situation of small farmers in the Andes, Bebbington highlighted the importance of peasant organizing to make it possible for them to sustain themselves and their production. He showed that when farmers joined together to negotiate with the state, international purchasers, or input providers, they had much more strength than if they were to take the same position as individuals.[10] And, the more mobilized they were, and the more experience they had in negotiating as a group, the more strength they would have.

Bebbington (1998, 1999) chose to speak of the potential of peasant organizing to ameliorate the negative effect of a number of factors, including the

---

10. See Getz (2003) for an interesting case study of this dynamic.

encroachment of the world market, in terms of social capital. Yet I would argue that what he described is the same protective reaction of society that Polanyi (1944) captured in the term *countermovement*.[11] That is, within this environmentally degraded and terribly impoverished region, small farmers responded to the expansion of market relations, with its corresponding social cost, by mobilizing to defend their lives and livelihoods in the way Polanyi's (1944) observations would have suggested.

McMichael (2006, 2008b) also places great emphasis on peasant organizing in the face of globalization. Writing of the need to reconceptualize "the agrarian question," given the changed nature of capitalism today—which has rendered nation-states obsolete as the organizing principle of political economy, their being replaced by global capital—he argues that peasant movements must now be considered a central part of it. But these movements, and he speaks of Vía Campesina as exemplifying this tendency, are not only seeking their own survival. In addition, they are setting forth an alternative vision of life (and of modernity) that redefines agriculture and their place in it. Their goal of food sovereignty—defined as "the right of each nation to maintain and develop its own capacity to produce its basic foods [in its own territory], respecting cultural and productive diversity"[12]— reflects their reaction to the "corporate-led process of agricultural commodity production" (McMichael 2008b, 212) that operates at the global level, as well as the need to reformulate the agrarian question so that it takes into account this new scale of political economy and the internationally organized peasants' reaction to it.

As will become evident, I share McMichael's (2006, 2008b) concern with highlighting peasant organizing in analyzing the agrarian question in the contemporary period. Recognizing the importance of social organizing by small farmers and their pursuit of multiple survival strategies introduces the notion of social agency on the part of these producers. It is not simply a matter of factors beyond their control that determine their continued existence. There are ways in which they, too, can play a role in avoiding their own elimination as a social category or class fraction.[13] But I differ with his

---

11. I find the concept of countermovement to be much less problem laden than that of social capital. For a critical discussion of the social capital concept, see Somers (2005).

12. Demarais (2007), as cited in McMichael (2008b, 210).

13. Edelman (1999) addresses this issue, and the multiple survival strategies pursued and organizational efforts engaged in, in the context of Costa Rica.

dismissal of the relevance of the nation-state in this equation. While I certainly acknowledge the heavy presence of the international economy and international capital in many social issues in the present era, including the peasant question, the case studies I present will show how crucial the actions of the nation-state continue to be even within the context of globalized neoliberalism.

In examining the cases of Nicaragua and Cuba, it will become clear that many of the above-mentioned factors—peasant organizing, pursuing multiple survival strategies, technological change, pressure from the international economy, and, perhaps most crucially, the intervention of the state—have played a role in defining the current status of small farmers in each country. The issue of the viability, or lack of it, of small-scale agricultural production (i.e., the classic agrarian question) has also risen to the fore in each country since 1990.

That year, both governments initiated an opening to the market. In Nicaragua, this opening formed a key piece of the more general retreat from socialism—or move toward the adoption of a capitalist orientation, to use Szelenyi and Kostello's (1996) language. In contrast, in Cuba, the shift was toward a reconfigured socialism in which the redistributive state was still hegemonic. Thus, these two countries represent ideal cases for exploring the extent to which varying degrees of incursion by market forces push small farmers toward extinction, while also highlighting the role the state plays in this process.

This study builds on others that have focused on single cases that are also located at the intersection of the theoretical debates around the social consequences of the transition from socialism and the "agrarian question" that this reignites. The Chilean case was, perhaps, the earliest one to allow for the examination of these issues simultaneously. Kay (1997, 2002) and others have closely analyzed the situation of small-holders since the 1973 coup, which brought an end to Chilean socialism and introduced neoliberalism, and found that they have lost much ground.[14] The first thing to hit many of these producers was the military government's counteragrarian reform, which returned a noteworthy portion of the land that had been expropriated during the 1964–73 period under the Christian Democratic government of Eduardo Frei and the socialist government of Salvador

---

14. See also Gwynne and Kay (1997), Murray (1997), and Gómez and Echenique (1998).

Allende to its former owners.[15] This left many former agrarian reform beneficiaries landless.[16]

The counterreform was followed by the implementation of SA policies
that further discriminated against the small farm sector. In a study of the
effect of agro-export booms on the rural poor in Chile, Guatemala, and
Paraguay, Carter, Barham, and Mesbah (1996) found that the government's
political economic policies had forced many of Chile's small farmers to sell
their parcels. At the same time, the growth in agro-export production had
given rise to greater labor absorption on the land, especially in the form of
seasonal employment. Yet between 1973 and 1989, real wages for agro-export
laborers had fallen below 1970 wage levels (Carter et al. 1996, 51). Although
wages rose somewhat after that time, as growing demand for laborers led to
shortages, the structure of the labor force also shifted from temporary workers being a minority to temporary workers being a majority (Kay 2002).
These findings indicate not only that many of those who had previously been
small farmers had been transformed into wage laborers with the implementation of Chile's counteragrarian reform and neoliberal economic policies
but also the absolute immiseration that must have characterized the lives of
agricultural workers who first saw their wages fall, then their opportunities
for full-time employment disappear, during this economic boom period.[17]

Others, whose work I will present in greater depth in Chapter 2 when I
discuss the cases of Russia and China, have likewise described the deepening
poverty and growing inequality in each of these countries as they opened up
their economies to the market. However, the studies of Russia (e.g., Hum-
phry 1998; Herrold 2002; Wegren 1998a) document the utter devastation that
has been wrought in the countryside by that country's SA, which was the
centerpiece of its retreat from socialism. In contrast, the well-being of Chi-

15. Belisario (2007a, 20) argues that the Pinochet government's counterreform was "partial"
in nature, in that not all of the properties that had been expropriated during the years of agrarian
reform were returned. Rather, approximately 33 percent were. Instead, the objective of the Pinochet government was to create a "modern" capitalist agricultural sector, which a complete counterreform would have only complicated. Hence, upper limits on farm size were left in place to
encourage a transition to "intensive" capitalist farming rather than a return to extensive landed
estates.

16. Still others who had benefited from the agrarian reform were excluded from the "parcelization" of the agrarian reform land that remained in peasant hands if they were unmarried, active
in left-wing political parties, were union organizers aligned with the Allende government, or for
a variety of other reasons (Belisario 2007b).

17. Kay (2002) also points to the replacement of a notable percentage of male workers by
female temporary workers as the nontraditional agro-export (NTAE) growers tapped into this
new source of cheap labor.

na's small farmers has varied, depending on the degree of market opening and the state's redistributive policies (cf. Travers 1985; Unger 2002; Riskin 1987; Riskin and Shi 2001). Thus, it improved during the first phase of the reform (1978–85), with the introduction of local markets within the larger context of a redistributive state (à la Szelenyi and Kostello 1996). With the further opening of the market from 1985 onwards—that is, the introduction of a socialist mixed economy—inequality grew within the peasantry. This was especially the case between regions, but it was also true within regions. Yet China's small farmers have not suffered the wholesale abandonment by the state that their Russian counterparts have. The situation of the Chinese peasantry is more nuanced, as well as more hopeful.

Finally, the experience of the peasantry in those Latin American countries that embarked on a comprehensive opening up to the market is also relevant here, even if it includes cases that never had completely redistributive states. Mexico represents an intermediate sort of case in this latter sense. An agrarian reform was initiated there in the revolutionary context of the 1930s (see Hamilton 1983), but its end point in the 1990s coincided with an opening up of the economy to the international market, most especially to the United States. Even during the period when state support for the peasantry in the reformed sector—the *ejidos*—had been relatively strong, indications of a slowdown in growth in this sector had begun to appear. Yet with the shift to market opening in the 1980s and 1990s, a full-blown crisis took hold in the *ejidatario* sector—which represented 75 percent of all agricultural production in Mexico (Davis 2000, 100)—and within the peasantry as a whole.[18]

State-supplied credit and other subsidies to agricultural production were cut, guaranteed prices came to an end, and the grain market experienced deregulation with the implementation of the North American Free Trade Agreement (NAFTA). Deregulation was meant to remove existing obstacles to the market unification that NAFTA gave birth to, thereby allowing for a flood of corn imports from the United States. That flood, which was only a rivulet of U.S. food exports, is part of what a number of scholars refer to as "food dumping" (see McMichael 2006, 2008a, 2008b; Araghi 2008). The term suggests the critique they are making of the unfair trade practices inherent in the subsidization of food production in the United States (and in

18. See also Lewis (2002), Gravel (2007), Assies (2008), and de Janvry, Gordillo, and Sadoulet (1997) for an array of accounts of the crisis, as well as of the response of the Mexican peasantry to it.

the European Union) that results in unsubsidized production in the Global South being unable to compete with it.[19]

As a consequence, Mexico's small-farm sector largely retreated into a subsistence mode of existence. The modification in laws pertaining to *ejido* property, which signaled the end of land redistribution and embodied the push toward privatization, only made easier the renting out and sale of that land by the population that had had its production totally undercut by the other SA measures. The result was the temporary and permanent alienation from that land—through rentals and sales, increased land concentration and social differentiation, and growing reliance on off-farm income, including from remittances from those who have migrated to the United States.

These patterns were even more accentuated where no significant redistribution had preceded the implementation of SA. In reviewing the history of a number of such countries, Carter and Barham (1996) found that, where landholdings were unequal before the policy shift, which fostered this opening, inequality deepened in its wake. The welfare of the smallest farmers worsened, while that of larger farmers improved. Bryceson (2000, 309) argued more generally that, as agricultural restructuring throughout the Global South took place, it "struck at the heart of the middle peasantries' agrarian base." That, in turn, produced "strong centrifugal forces of economic polarization and class differentiation set in." What these scholars described was much like the process that Lenin and Polanyi, as well as Szelenyi and Kostello many years later, had predicted would occur with the spread of capitalism.

Conclusion

This brief foray into these several areas of debate has pointed to a number of patterns that the case studies and the comparison of them that follows will speak to. First, they underline the strong relationship between expansions in market relations—as described for the liberal era by Polanyi (1944), the movement away from orthodox socialism toward the market embodied in socialist mixed economies and capitalist economies analyzed by Szelenyi and

---

19. Others (such as Kay 2006) also describe this phenomenon and its implications for small farmers in the Global South, but without using the term *dumping*. McMichael (2008b, 211) states that "the critique of dumping . . . illuminates . . . the institutional construction of a corporate market [with agri-business at its core] premised on 'naturalizing' peasant redundancy, through political means."

Kostello (1996), and where neoliberalism was imposed on capitalist economies that had previously been protected to some degree—and increased inequality. For Polanyi (1944), this "movement" was highly destructive for the livelihoods and lifeways of England's poorer folk. Those writing about more recent periods reached similar conclusions about their regions of study.

Central among those poorer folk were the peasantry. While not seeking to speak directly to the "agrarian debate," Polanyi's (1944) depiction of the effect of "the movement" coincided closely with that of a number of scholars whose work focuses specifically on it. Whether or not those participating within this debate had precisely the same dire predictions as Engels (1977), Lenin (1957), and Polanyi (1944) had expressed about the future of the peasantry with the expansion of capitalism into agriculture, by and large they characterized the lives of the peasantry as having been made more difficult because of it.

Yet, Polanyi (1944) also argued that "countermovements" typically emerged that reacted against those "movements," seeking to ameliorate their consequences. Here, too, we see resemblances to this dynamic in many of the latter-day analyses of the agrarian question. Regardless of whether those countermovements have taken the form of actual organizing by the peasantry, their pursuit of diverse survival strategies, or state intervention on their behalf, many of these scholars have implicitly or explicitly revealed the "double movement" foreseen by Polanyi (1944).

With this short theoretical introduction completed, it is time now to look more closely at the social consequences of the Russian "retreat from socialism," and the Chinese "reconfiguring of socialism." Examination of them will facilitate an understanding of the significance of what is described in Chapters 3–6, where my findings on Nicaragua and Cuba are presented. By looking first, however superficially, at the "transitions" that have occurred in Russia and China, the broader implications of Nicaragua's and Cuba's distinct pathways toward the market will be clearer.

2

Small Farmers in a Contrasting Light

Nicaragua and Cuba represented two of the smaller socialist regimes that had come to power. Yet the changes they experienced as they went their separate ways in 1990 paralleled, in many respects, the changes that were under way in two countries that had been the world's largest socialist regimes, Russia and China. The last quarter of the twentieth century was a time of major transformations in Russia and China. Following more than seven decades of "building socialism" in Russia and almost three decades in China, the state in each of these countries shifted its orientation significantly.

In Russia, the largely pacific overthrow of the socialist regime opened the way for a pro-Western government to begin a rapid retreat from socialism. That shift reached into virtually every area of policy making, not the least of which was agriculture. Hence, policy reflected the objectives of dismantling socialist agriculture and replacing it with the newly established capitalist agriculture that the post-Soviet "agrarian reform" was to give rise to.

In contrast, China's shift in orientation entailed a reconfiguration of socialism to be brought about through a new era of reform. Although it would involve dramatic changes in both the economy and society, this reform was not aimed at leaving socialism behind. Rather, it was geared toward making a variety of modifications within it to overcome the obstacles to further growth and overall political and economic viability that had emerged in the final years of Mao Zedong's rule. Here, too, agriculture would experience noteworthy changes. But these changes were designed to improve the production and livelihoods of the peasantry, who were its principal protagonists.

Thus, the state in these two socialist giants opted to move in distinct directions from each other. As a consequence, their new projects had quite different implications for their respective agricultural populations. Let us now examine their varying approaches to the market, thereby laying the foundation for understanding the ways in which their trajectories were similar to, and different from, those of their smaller counterparts.

Post-Soviet Russia

Russia's process of transformation began later than China's. But once it was initiated, the modifications set in motion were even more far reaching. After half a decade of relatively modest alterations in the Union of Soviet Socialist Republics' economy under Mikhail Gorbachev, reforms that with time might have put the USSR on a course similar to post-1978 China, the socialist regime in power completely disintegrated. So began that country's race to retreat from socialism.

With Boris Yeltsin's rise to power in 1991, the most radical reformers came to the fore in the USSR's final days. Whereas Gorbachev had substantially opened up the USSR politically during the later 1980s, his approach to economic reform was much more cautious. With Yeltsin, caution was thrown to the wind, and Western advisors who promoted a shock therapy for the country's serious economic ills were brought in to guide him. Their approach was of a piece with Yeltsin's political economic vision for Russia— one of Western-style capitalism. Their proposed therapy was designed to move Russia as quickly as possible toward this goal (see Nolan 1996/1997).

The policies that composed this shock therapy included, among others, an instantaneous price liberalization, imposition of austerity measures, legislation of a very rapid process of privatization, and lifting trade barriers. All of these policies have had an effect on agricultural production and those who engage in it. Perhaps the most overarching of them, in terms of creating the conditions within which agricultural production has been carried out, has been privatization.

Before the collapse of the USSR, agricultural production there had been primarily large scale. It had taken place on state farms (*sovhozes*) and collectives (*kolkhozes*). These farms had been formed through the agrarian reforms and collectivization drives of the period between 1917 and 1930. Their average size became bigger over time as collectives were fused together into ever larger groupings. By 1988, the average-sized collective had 6,300 hectares and 954 members, while the average-sized state farm had 15,600 hectares and 1,082 workers (Laird 1997, 472). Given the large scale of their production, higher and higher levels of technology had been introduced to the state and collective farm sectors over the seventy years of Soviet rule.

However, there was an additional sector of agricultural production—the personal plot. In principle, state farmworkers, collective members, and non-agricultural workers could farm personal plots. But most commonly, collective members worked the tiny plots of land.[1] Rudimentary levels of technology were the norm. Yet production there was actually quite important in terms of the food products it made available to the families who farmed them and the large population that purchased their surplus, as well as its contribution to these farmers' incomes.

Nonetheless, Soviet agriculture was characterized by tremendous inefficiencies. By the 1980s, growth in agricultural production was not keeping pace with rising investments in this sector nor generating commensurate increases in consumption, and problems of adequate incentives for state farmworkers and collective members continued to perplex policy makers (Wegren 1998a; Nove 1986). Nickolsky (1998, 191) points to the imbalance that existed between agriculture and industry in the USSR as the principal cause of these inefficiencies. The "Soviet obsession with industrialization" (Nickolsky 1998, 191) led to the siphoning of resources from agriculture into industry, impoverishing the peasantry and the countryside in the process.

Soviet leaders had been aware of the weaknesses in agricultural policy since at least the 1950s. Each successive administration had attempted to address these weaknesses through a distinct combination of policy measures. Gorbachev had gone the farthest in terms of combining organizational reform, support for production on the personal plot, and investing in agriculture. Yet on the eve of the Soviet Union's collapse, food shortages and other by-products of the sector's inefficiencies still represented challenges for those concerned with agrarian policy.

Within this context, the new political economic orientation was adopted. One of the first legislative measures passed by the recently formed Russian

---

1. According to Wegren (1998a, 33), despite some regional variation, overall personal plot size, including the land the family's home was built on, could not exceed half a hectare.

government in 1991 was a presidential decree designed to bring about a radical change in agricultural property relations. According to one analyst (Lerman 2002, 43), "This was a systemic sectoral reform, whose main goal was to transform the agricultural sector into a system compatible with market-oriented economic principles that were part of the overall transition to the market."[2] More specifically, this decree stipulated that all state and collective farms had to be reorganized during 1992 into joint stock companies, limited liability companies, or other kinds of private enterprises. But the decree stopped short of allowing the sale of the newly privatized land.[3]

Despite the relatively limited nature of this measure, its implementation moved at a much slower pace than its authors had envisioned. After two full years of agrarian reform, 34 percent of these farms had re-registered as *kolkhozes* or *sovhozes*, while another 59 percent had registered as joint stock companies (Nickolsky 1998, 199). Follow-up legislation, in the form of another presidential decree in late 1993, mandated that all former state and collective farm property be divided among workers or members. This was to take place through the issuing of certificates that, when grouped together, were equal in value to the property in question. The second decree allowed for the certificates to be repooled and held collectively or to be subdivided to individuals to be farmed individually or in joint stock companies.

Nevertheless, most former state farmworkers and collective members continued to work the land together, having chosen to pool their certificates. Thus, in 1994, large farms (including the "personal plots" located on their land) accounted for 95 percent of the country's cultivated area and 98 percent of its output (Kitching 1998a, 19).[4] These figures remained almost the same into the new century (Ioffe, Nefedova, and Zaslavsky 2006, 35–36). The much-heralded "peasant farms" to be formed by the mass of rural dwellers who were to become the basis of Russia's new rural capitalism remained only a footnote in the history of post-Soviet agriculture. Although peasant farms occupied roughly 5–6 percent of the country's agricultural land, they pro-

---

2. Wegren (2004, 558) argues that "Yeltsin's rural policies were explicitly intended to destroy Soviet-era rural institutions that were perceived as bastions of the Communist system."

3. Gambold Miller and Heady (2003, 257) suggest that the law was limited in this sense out of "the desire to keep ownership within the local community and to protect country dwellers from unscrupulous land speculators."

4. Laird (1997, 475) coincides with Kitching in noting that peasant farms accounted for only 6 percent of plowed land and 5 percent of all agricultural land. In contrast, Lerman (2002, 51) includes the personal plots with the individual family farms in arguing that "the individual sector today controls 10 percent of agricultural land in Russia, up from 1.5 percent prior to 1990."

duced 2 percent of output in the mid-1990s,[5] and their share of output had
gone up only slightly by 2001 (Ioffe et al. 2006, 35–36). Because of the difficult economic circumstances faced by all farmers, but especially individual
farmers, it was likely that their presence would continue to be limited.[6]

From the point of view of the reform's major promoters, the less than
total enthusiasm that the privatization process encountered in the countryside was troubling. Not surprisingly, given the overarching assumptions behind their economic policies, they associated the massive drop in production
that took place at this time (to be described later) with it. And in 2002, they
succeeded in pushing a new law through the legislature that went that final
step toward making land fully private—allowing for its sale by those who
received it under the 1991 and 1993 privatization laws. However, to date, the
market for land leasing has been significantly more active than for land sales
(Wegren 2008). The latter has been restrained by local government restrictions, as well as by the resistance of local elites and the rural rank and file
(Pallot and Nefedova 2007; Sazonov and Sazonova 2005).

Why was it that the "beneficiaries" of Yeltsin's agrarian reform opted to
continue to hold their land collectively throughout the 1990s and into the
new century despite the Soviet history of forced collectivization? The answer
was that even though the "command economy" that had characterized the
Soviet Union had disappeared, along with the requirement that farms fulfill
production quotas and depend entirely on the state for all inputs, individual
farmers did not have much chance of survival because of the economic environment. Rather, relying on the multiple connections that former state farm
and cooperative administrators had to obtain production inputs, as well as
their provision of social services (albeit now greatly diminished), offered
more hope for most of the rural population than "going it alone" as private
farmers.

Although the reform administrations that have held office since the dissolution of the USSR had the construction of capitalism as their principal
objective, in agriculture, the only real step taken to facilitate this was the
privatization of state and collective sector farms. As students of agrarian
reform are well aware, the receipt of land is not enough to put agrarian

5. The first figure is an estimate based on figures provided in Lerman (2002, 51) and Wegren
(1998b, 83). The latter figure is taken from Lerman (2002, 51).

6. Kitching (1998a, 3) went so far as to say that they are "leaving the land in droves." And,
Laird (1997, 475) noted a drop of 2 percent in the number of private farms in the first nine months
of 1995 alone.

reform beneficiaries on firm footing anywhere in the world. Credit, technical assistance, rural infrastructure, and marketing assistance are usually necessary complements to land if the reform is to really take hold. Yet in the case of Russia, the privatization of land was carried out in the context of a structural adjustment (SA), which severely undercut the possibilities for taking such an integrated approach.

Ironically, most of these other resources were more accessible before the agrarian reform of the 1990s than after it. For example, agricultural credit had been readily available within the Soviet system of agriculture. This resource is important for funding production-related expenditures during the relatively long period before the returns begin to come in following the harvest. In the USSR, the "soft" nature of credit meant that state budget deficits increased. Aside from the state's systemic-level problem, credit basically eliminated the need for cash in the Soviet economy. Producers were granted credit notes (not cash) to purchase all of their inputs, and the state procurement agencies set aside part of the payment for the crops they purchased as repayment for the credit extended at the beginning of the agricultural cycle. Even though little cash flowed through the system, lack of credit was not an obstacle for production.

However, the credit situation changed drastically in the post-Soviet era. The government had promised loans at low interest rates to incentivize private farming, but its SA measures precluded its fulfillment of this promise. Thus, state-supplied credit was a tiny fraction of what it had been before 1990. Describing the situation in the first half of the 1990s, Laird (1997, 474) said "virtually none [of the individual farmers] have the cash or the credit lines necessary to purchase . . . needed inputs."[7] This situation was also true for those who had opted to maintain their production in a collective form. The exorbitant interest rates charged by private banks made their credit unattainable for most producers. A few more options opened up for farmers in the second half of the 1990s, including the development of rural credit cooperatives starting in 1996. According to a 2001 rural survey, though, the number of credit recipients was only 6 percent (Wegren 2004, 567). Perhaps, with the opening of a pilot program announced by the state agricultural bank in 2003 to supply household plot owners with credit, this resource would become more available. In the meantime, the agrarian economy had, as of 1991, become credit starved and largely cash based.

7. Also see Wegren (1998a) on this point.

Nevertheless, as was true for the other sectors of the economy, cash was also in extremely short supply. To "purchase" inputs, producers engaged in a multiplicity of barter arrangements. Inputs were then exchanged for a portion of the potential harvest. These barter arrangements were often quite complicated, involving whole chains of exchange agreements.[8] And, they depended on the producer having a wide range of connections for their construction. This was one of the key senses in which those who farmed in a collective fashion were at an advantage over those who farmed as individuals. The former were represented by administrators, many of whom were in the same respective positions in Soviet times. Therefore, they had a wide range of contacts spanning many years. They could also dedicate themselves full time to ensuring the functioning of these barter arrangements (much as they worked to ensure the timely arrival of inputs in the former era).

Yet aside from the challenge of establishing and maintaining barter networks, there was the additional issue that the price of inputs had risen notably during the 1990s in relation to producer prices. This is well illustrated in the following figures: "from 1991 through 1995 prices for industrial goods used by food producers increased 2,230 percent, while purchase prices for agricultural products rose only 752 per cent" (Wegren 1998b, 95).[9] This problematic "price scissors" situation prevailed into the new century (see Ioffe et al. 2006). For Humphrey (1998), the detrimental terms of trade were linked with the lack of credit, in that it gave those controlling access to inputs—the "Products Corporation," which was a state enterprise in the case she describes—a special advantage over those who needed to purchase them and then were required to sell their produce to inputs suppliers at prices established by the suppliers.[10] Thus, it took either great resources or tremendous barter relations to keep agricultural production going in the post-Soviet period.

Many things combined to weaken the value of agricultural produce during this period. For example, many of the subsidies that had previously underwritten Soviet agriculture were eliminated with the implementation of SA because they were seen as impeding the free functioning of the market. Demand for produce also fell after 1990, as the population's purchasing

8. See Kitching (1998a) and Humphrey (1998) for accounts of bartering chains producers relied on to obtain inputs.

9. Nickolsky (1998, 207) actually says that this gap increased 4.3 times in favor of industry. See Allina-Pisano (2002) on the effect on the ground of deteriorating terms of trade.

10. Wegren (1998b) also speaks of the damaging effect on prices of state monopolies being the principal purchasers of locally produced food.

power was likewise reduced by Russia's economic reform. Consumer subsidies were eliminated, starting in 1992, and other kinds of subsidies also disappeared. This occurred at the same time as receipt of wages became increasingly erratic, undercutting consumer capacity to purchase food accordingly.

From the agricultural producers' viewpoint, the situation was further complicated by food and other imports that flowed into Russia with the liberalization of trade in the post-1990 era. Trade liberalization is a centerpiece of SA and is heavily promoted by advocates of market economics. Nonetheless, its effects vary for different sectors of the economy, with locally oriented producers typically not faring well.

Russia's trade liberalization brought about a drastic reduction in tariff levels. By the mid- to late 1990s, Russian tariff levels for most products were only about one-half of world averages (Wegren 1998b, 100). Hence, food imports increased significantly over the 1990s, reaching the point that they "comprised over 25 percent of all imports in 1996—the largest single category of imports." Consumers, to the extent that they had purchasing power at all, preferred to use it for the higher-quality and less expensive imported goods.[11] Russia's producers could only lower their prices so far—and clearly not far enough to compete with imports—because of the high cost of their inputs and the multiple adversities they confronted in their production, including the less than ideal ecological conditions (e.g., soil quality and climate) that characterize the country. The result was that their position in the local produce marketplace was extremely weak. They were price-takers, not price-setters, and price-takers of very poor prices, indeed.

The 1998 financial crisis, though, modified this situation somewhat. As it brought with it a devaluation of the local currency, imports became more expensive. The government also began to put in place import quotas on certain specific goods, such as raw sugar, chicken, and red meat, thereby protecting producers of those goods to some extent. But this was only after agricultural producers of all sizes had been struck hard by the government's agrarian policies for almost a decade, leaving them in a greatly weakened position.

Given the many difficulties and uncertainties confronting Russian agriculture, it should come as no surprise that production dropped massively

---

11. This situation prevailed for other items as well, as Humphrey (1998) describes for locally produced versus imported wools.

after 1990. By early 1995, official statistics acknowledged a 33 percent reduction in gross food output since 1990 (Wegren 1998a, 127). Overall food production continued to fall in subsequent years: "In 1998, the output of crop farming was just 56 percent that of 1990 . . . and the relative output of animal husbandry was only 49.7 percent" (Ioffe et al. 2006, 28). The fall in production was particularly dramatic among Russia's key crops: between 1990 and 1995, grain production decreased by 46 percent, meat production by 42 percent, egg and milk production by 29 percent, and sugar beet production by 39 percent (Wegren 1998a, 128). Production drops were both absolute, in terms of concrete amounts of produce being less than in the past, and relative, in terms of notably lower yields. For example, between 1990 and 1995, grain yields decreased by 37 percent, and milk, egg, and meat yields also fell (Wegren 1998a, 127). This situation characterized Russia's more productive, as well as less productive, regions.[12] Some improvements occurred in the wake of the financial crisis and the implementation of more favorable agricultural policies under Putin (cf. Wegren 2004). Yet "output in 2002 was still only 67.5 percent of what it had been twelve years earlier" (Ioffe et al. 2006, 28).

Aside from what has already been described, a number of other reform-related factors contributed to this significant drop in production. Investments, particularly on the part of the state, had fallen precipitously during this period. According to Nickolsky (1998, 195), government investments in agriculture in 1993 were only 6 percent of what they had been during the 1986–90 period.[13] This alone was bound to have a major effect on the sector, but other issues also came into play. Because of the lack of access to credit and cash, application of agricultural inputs fell. In 1994, application of mineral fertilizers was only 11 percent of what it had been between 1986 and 1990, and the comparable figure for organic fertilizers was 29 percent (Nickolsky 1998, 207). Although at first glance this might appear to have represented a step in the direction of more sustainable agriculture, given Russia's generally poor soils, it produced increased land erosion and decreased soil fertility. This, in turn, would have a major ongoing effect on yield levels.

Yield levels were also affected by falling labor productivity. Perhaps the principal cause of this was the not so gradual process of demechanization

---

12. Wegren (1998a) and Allina-Pisano (2002).

13. According to Ioffe, Nefedova, and Zaslavsky (2006, 28), "from 1965 to 1985, Russian agriculture was receiving 28 percent of the total investment in the Russian economy. In contrast to that, in 2001 agriculture got just 2.7 percent of the vastly diminished total investment."

taking place in the Russian countryside. Although Soviet levels of mechanization were low by U.S. standards, the situation deteriorated notably after
1990. According to Wegren (1998b, 103), "only about 70 percent of farm
machinery is operational and large farms have neither money to buy new
machinery nor repair the old." Moreover, Laird (1997) states that not only
the quantity but also the quality of farm equipment declined. Because of
ever-present problems of spare-parts shortages, the working life of existing
equipment was substantially less than in other industrialized countries. The
most immediate cause of the demechanization of Russian agriculture was
the lack of available credit and the overall problematic economic circumstances in the countryside. Sazonov and Sazonova (2005) argue that farms
were forced to sell their machinery to improve their cash flows, precisely
because of the absence of credit. The result was reduced productivity of labor
on larger farms, where the need for machinery was greatest. This translated
into reduced yield levels and falling production.

Falling production had a number of important consequences. One was
the need to import food. Nickolsky (1998, 207) argues that declining productivity and production caused a dramatic increase in food imports. In 1995, 35
percent of the food sold in Russia was imported. This was so even as lack of
purchasing power reduced consumption of marketed agricultural goods.[14]
Yet, as mentioned, food imports further weakened incentives for local production.

Falling production, in combination with lack of access to state credits, and
deteriorating terms of trade, also wreaked havoc on the financial situation
of Russia's agricultural producers. Wegren (1998b, 102) notes that financial
insolvency of the collective farms grew from characterizing 10 percent of all
such enterprises in 1993 to characterizing 80 percent in 1997. By early into
the new century, that figure was between 83 and 86 percent (Ioffe et al. 2006,
36). These figures alone illustrate the extent of the economic crisis in Russia's
countryside. Producers were barely holding on by a thread, and that they
had been able to hold on at all was only because there was no institutional
structure to impose foreclosure on them and no one to buy up their farms.[15]

Even as Russia's traditional agriculturalists attempted to hold on, they
had to retreat further into subsistence production. This can be seen most

14. On declining food consumption, see Laird (1997) and Wegren (1998b).
15. Although, increasingly (as will be described later), their members' shares were being
leased to the "New Agricultural Operators" who represented the most vibrant, capitalistically
oriented part of the economy.

clearly in that the drops in production were, by and large, concentrated in the large farm sector. Thus, while production fell precipitously on the acreage that had previously formed the state and collective farms, output actually grew by 6 percent in the private plot sector (Kitching 1998b, 52). The large farms continued to maintain their dominance in grain crops. However, in essentially all other crops and livestock products, the private plots were more important.

The relative rise of the family, or private, plot was an outcome of the reform policies instituted after 1990. In the past, the *kolkhozes* and *sovhozes* basically ensured the well-being of their members/workers from birth to death. Nevertheless, with the implementation of SA and the agrarian reform, these collective structures were ordered to relinquish their social welfare functions and to operate as nothing more than productive structures. Although this change was not fully realized, most collectives were no longer in a position to maintain the level and array of services they provided in the Soviet era, and the void created had yet to be filled by another institution.[16]

At the same time, the serious cash shortages described earlier had an effect on the collectives' ability to pay cash wages. On those collectives that still paid cash wages—a considerable minority did not at all—their receipt was often infrequent and at reduced levels from what they were in the past. To the extent that the collectives were functional, their members were often paid in-kind. In-kind wages typically took the form of providing machinery services to their members with private plots and hay and other feed crops for the livestock of their members. Given the small size of the private plot and the importance of livestock for the collectives' members, the receipt of feed was, indeed, significant for them.

As essential, though, was the members' continued access to their private plot. Production there ensured their and their family's food supply. Moreover, because of the cash shortages experienced throughout the Russian economy, even urban-based relatives came, increasingly, to count on what could be produced on these "private plots."[17] In some cases, this meant relying on their rural-dwelling relatives for farm produce. Hence, entire extended families, effectively, depended on the collective. In other cases, urban dwellers gained access to plots that they worked on weekends. According to

16. In fact, a number of their social welfare services, such as the provision of health care and education, were supposed to pass to local government institutions, which were desperately underfunded.

17. See further Burawoy (2000) and Humphrey (1998).

Zavisca (2003, 786), "nearly half of urban Russian households grow food on [such] dacha plots."

The expanding role of the private plot can be appreciated through a glance at its weight in the production of certain key food products. For example, by 2004, private plots produced more than 52 percent of all milk (up from almost 39 percent in 1994 and 28 percent in 1990), more than 52 percent of all meat (up from 43 percent in 1994 and 30 percent in 1990), an overwhelming 80 percent of all vegetables (this figure was 67 percent in 1994 and 33 percent in 1990), and almost 92 percent of all potatoes (as opposed to 88 percent in 1994 and 65 percent in 1990).[18] Clearly, without its contribution to production, untold millions of Russians would go hungry.

This strong reliance on the collective for assistance with private plot production, in terms of inputs and services, was central in its members' commitment to keeping the organization intact. Some analysts suggest that this was the only thing tying members to the collective (e.g., Kitching 1998b; Allina-Pisano 2002; Pallot and Nefedova 2007). Yet Humphrey (1998) argues that, in addition to the dire need for the inputs that the collectives provided, there continued to be a sense of community in them that originated in the distant past and that would not be let go of easily. Furthermore, the collectives represented a "mini-state" in the face of the failure of the Russian state to provide any sense of order and security.[19] Whatever the differing weight of these factors in the decision of former members to throw in their lot together, the result was a recasting of the old collective model to aid in survival during the new era.

As the financial crisis of 1998 began to pass, and Putin's policies began to lighten the burden borne by agriculture since 1990, a new group of actors emerged to take advantage of opportunities that existed in the agrarian sector. They have come to be known as the "new agricultural operators" (NAOs). By and large, the NAOs are outsiders to the rural sector who lease "shares" from the collectives to gain access to this key factor of production. Rylko and Jolly (2005, 116) note that they have a number of characteristics in common: "They are exceptionally large, . . . externally owned and managed, commercial farming operations." Some of them were food-processing enter-

18. Data for 1990 are taken from Wegren (1998a, 45); for 1994, from Kitching (1998a, 21); and for 2004, from Pallot and Nefedova (2007, 18).

19. Yet this sense of the collective filling in the vacuum left by the state had been largely lost by the middle of the next decade, and members increasingly felt they could only rely on themselves, their family, and their neighbors for assistance that had previously come from these institutions (see Velikii and Morekhina 2006).

prises that have gone into agricultural production to ensure the crucial agricultural inputs they need to keep their industrial operations functioning. Others are simply venture capitalists from other sectors entirely (e.g., oil and gas) looking for areas to invest in. While their access to credit has enabled them to make significant investments in their production in this sector, their bottom line—generating profits—has led to layoffs on the lands they lease. Hence, their presence in the rural landscape has not greatly improved the position of the mass of small farmers.

Instead, the reliance of that mass of small farmers on their private plots has only increased in the interim. Nonetheless, having access to food and some income from private plot production only slowed the decline into poverty that much of the rural population experienced following the imposition of the agrarian reform and SA. According to von Braun, Qaim, and Seeth (2000, 307), 75 percent of the population fell below the poverty line based on the cost of living and income levels, and this figure reached 82.5 percent in rural areas. However, if one takes into account other factors, such as caloric intake, this figure dropped to just over 19 percent at the national level (with no significant differences between urban and rural areas). Although figures are unavailable for poverty levels before 1990, students of this phenomenon argue that the social safety net that existed in Soviet times succeeded in keeping the population above the poverty line (von Braun et al. 2000). Furthermore, as of 1992, the four-decade policy of trying to eliminate rural-urban wage differentials was reversed. The differentials went from state farmworkers earning 105 percent of the wages of industrial workers, and collective farm members 88 percent in 1990, to state farmworkers earning 63 percent and collective farm members 49 percent of the average industrial worker's wages in 1992 (Wegren 1998a, 124–25). This pattern has only continued since that time.[20]

Added to this was growing rural unemployment. In speaking of the ending of the social contract that had been implicit in collectives and state farms before 1990, Humphrey (1998, 452) states that "in 1992–93 the newly voted-in directors in all types of collective took the opportunity to economize by divesting themselves of 'useless' workers." Wegren (1998a, 234) notes a 370 percent rise in rural unemployment between 1992 and 1994, with this figure doubling during 1994 alone. Ioffe et al. (2006, 91) suggest that by 1999 overall rural unemployment was approximately 18 percent.

20. Bogdanovskii (2005, 148) states that "by 2002 [the monthly nominal wage paid by farm enterprises] had fallen to 40% of the national average and 35% of the average industrial wage."

These various dynamics led Wegren, Patsiorkovski, and O'Brien (2006) to conclude that a new class structure is emerging in rural Russia. Whereas the reform had initially produced differentiation and stratification in the Russian countryside, these processes have more recently moved beyond this to give rise to a novel class structure. They only distinguish between well-off and poor households. But they argue that, in addition to income stratification, distinctions between these two groups can be seen in landholdings, capital stock, class consciousness, and shared attitudes and values (Wegren et al., 379). This represents a shift from the rural class structure in the Soviet era, which included collective farmworkers, state farmworkers, individual peasants, and others. Despite these class distinctions, Wegren et al. (2006, 378) argue that "by the mid-1980s the Soviet countryside was not completely equal, but it would be misleading to portray rural social structure as divided into [the categories of] poor, middle, [and] rich." Although these scholars are reluctant to speak of an intermediate social category in the present period, their schema points to definite groups (or classes, as they designate them) of haves and have-nots. And the development of this new class structure is a product of the process of market reform.

Moreover, the market reform also produced a dramatic fall in investments in rural amenities, services, and facilities (such as schools, medical care, etc.), which only made life for Russia's rural have-nots more difficult after 1990. In addition to poverty figures, this could be seen in the decline in birthrate and increase in mortality rate in rural areas (see Wegren 1998a, 236). It was also evident in the exodus from farming that occurred after 1990. Between 1991 and 1995, the number of farmworkers involved in agricultural production dropped by 26 percent (Wegren 1998a, 126).[21]

This trend was especially strong on the large farms where, by 2002, the number employed was only 40 percent of what it had been in 1990 (Bogdanovskii 2005, 143). Those most likely to leave were younger workers, who saw little future in farming. Thus, the farm population experienced a rapid "aging," further undermining its prospects for the coming years. The grim situation in the countryside was not lost on those who were still too young to enter the workforce. In a series of surveys of rural youth attending high school and an agrarian university, most of the latter group said they planned to live and work "in the city" after graduation, and more than half of the

---

21. Herrold (2002) describes at length the rural exodus, especially of young adults in her area of research in Siberia.

former group said the same (Kazakbaev 2006, 72). So, a reversal of this pattern in the near future did not appear likely.

Clearly, Russia's transition to the market, although far from complete, had had a strong effect on its agricultural producers, especially those who had formed part of the state farm and collective sectors. The implementation of SA there—as embodied in the drive to privatize agricultural resources, the austerity measures that curtailed subsidies, agricultural credit, and investments of all types in the sector, and the trade liberalization that undercut local producer prices and contributed to the worsening terms of trade for agricultural produce—had disadvantaged small farmers in a multitude of ways. Among the consequences of these disadvantages was their deepening impoverishment. At the same time, the number of people dependent on their small plots for sustenance had grown. In sum, Russia's embracing of capitalism increased the level of inequality existing within society and severely weakened the position of the majority of agricultural producers, particularly those who had represented the backbone of Soviet agriculture: state farmworkers and collective members.

But, the transition to the market did not bring about the marginalization of the peasantry everywhere it was set in motion. The path chosen by those guiding this process was crucial in determining whether inequality would increase, as well as who would be the winners and who would be the losers. The contrast with Russia inherent in my second comparative case—China—illustrates well these important political economic dynamics.

China

Shortly after the death of Mao Zedong in 1976 and the arrest of his ultraleftist followers—the Gang of Four—who had, effectively, presided over policy making for the previous ten years, a new era of reform was inaugurated in China. The intense political struggle that had led to their arrest reflected the tremendous political upheaval set in motion by the "Cultural Revolution" the Four had led.[22] A desire to bring an end to that upheaval contributed heavily to the reforms.

However, economic factors also carried weight in the decision to embark on the reforms. Despite impressive advances in some areas of the economy

22. See further Nolan (1995) on the political costs of the Cultural Revolution.

during the preceding decade, disequilibria had emerged in crucial sectors of it. As in the case of the USSR (as well as many industrializing countries more generally), imbalances existed between industry and agriculture. The latter sector was squeezed, especially during the 1950s, so that the surplus produced there could be used to fuel China's industrialization. Within agriculture, technological improvements—in terms of employment of farm machinery and chemical inputs—had been notable (see Perry and Wong 1985), and yet growth in output had not been commensurate. When this was combined with rapid population growth in rural areas, the outcome was a drop in produce marketed. This resulted in urban food supplies diminishing, which led to increased food imports. In addition, a lack of rural off-farm employment produced low per capita income growth in the countryside. In the industrial sector, serious imbalances existed between heavy industry and supporting industries leading to problems in supply and demand. At the same time, many state-run enterprises were operating at a loss. Wage freezes in this sector had also led to notably low worker morale, with its consequent effect on production.[23] Clearly, some changes in economic policy were in order.

Mao's death created the opening needed for reform-minded policy makers to step up to the helm. After a brief period of relatively modest reform under Hua Guofeng, Deng Xiaoping took over the lead with a commitment to set in motion a reorientation of policy away from the Maoist legacy.

Although the reforms carried out in China since 1978, in many senses, have been radical, the approach to those reforms differed significantly from that taken in post-Soviet Russia. Whereas in the latter case, speed was obviously a priority, in China a much more incrementalist tack was taken. It depended heavily on local experimentation of alternative policies, which was studied closely at the central level before receiving approval for more large-scale adoption. And, instead of dramatically curtailing the role of the state in the economy—as occurred in post-1990 Russia—the Chinese state maintained its posture as a "developmental state" seeking to foster economic growth while raising the country out of poverty. Moreover, sustaining China's independence continued to be a priority. A strong planning apparatus and substantial public sector were to ensure that these objectives were met. Thus, what the Chinese state embarked on in 1978 was the reconfiguration of

---

23. Sato (2003, 178), likewise, points to stagnating wages on the agricultural communes as leading to a lack of incentives among their members.

that country's socialist model of development to achieve its "modernization," rather than a full-scale retreat from socialism.

But, what exactly did that reconfiguration of socialism consist of? It consisted of important changes in a number of areas of policy making. The changes that most non-China specialists are aware of took place in the externally oriented sector of the economy. China actually began to come out of its period of isolationism and autarky during the Cultural Revolution, with its efforts to reestablish relations with the West. With the initiation of the reforms, these efforts were broadened. For example, the 1979 Law on Chinese-Foreign Joint Ventures sought to encourage foreign investment. Typically, such efforts are motivated by a need for foreign capital. In China, though, the idea was to assist the country in catching up on the technological front. As part of the overall reconfiguration effort, joint venture firms were to be structured so that they would stimulate many backward linkages into the local economy. Foreign investment was not supposed to follow the common path of creating enclaves in which foreign capital operated and from which profits were expatriated unrelated to what was happening in the rest of the economy. Instead, with the state's strong "guidance," investment would become a channel through which information about the international marketplace would flow.[24] At the same time, the state imposed some heavily protectionist measures that were designed to keep certain new industries closed to foreign penetration. Most of the joint venture firms were located in the industrial sector of the economy, and their production was, by and large, geared for export.

In the agricultural sector, the changes were, perhaps, even more remarkable. In an attempt to redress relatively stagnant production levels, a number of policies were modified. The most dramatic of these modifications was the downsizing of agricultural production. Following a period of land redistribution that coincided with the victory of Mao and his army in 1949,[25] what began as a gradual process of increasing levels of cooperation between farmers was greatly accelerated in the 1955–58 period. This culminated in the disastrous Great Leap Forward campaign of 1958, the consequences of which subsequently caused state policy makers to backtrack a bit on their collectiv-

24. Nolan (1995) describes several instances in which multinational firms had to agree to assist in the development of local productive capacity of associated industries to be able to establish their joint venture firms.

25. This redistribution took place earlier in some places than others, depending on the level of support held by the Chinese Communist Party (CCP) in any given area and the consequent consolidation of its power.

ization efforts and to return the scale of production to a smaller level. Yet, by 1978, large-scale communes once again dominated the rural landscape. These communes averaged 2,033 hectares, typically with a membership in the neighborhood of 3,346 households (Oi 1989, 5). The "bigger is better" philosophy was clearly hegemonic within agriculture.

Nonetheless, within this general framework, small family plots were permitted. According to Riskin (1987, 27), during the decade preceding the reform, family plots covered an average of 5–7 percent of the cultivated land.[26] Despite their diminutive size, as in the Soviet case, the plots were important for those who farmed them. They provided farmers with the opportunity to produce meat and vegetables (as opposed to the grains that the cooperatives specialized in), as well as the possibility of generating cash income. That income was derived from the sales, whether legal or otherwise, of their surplus production to town dwellers.

The agricultural means of production, including land, work animals, and machinery, were collectively owned at the production team level (a subdivision of the commune, which averaged thirty-three households), while industrial capacity was collectively owned at the brigade (the tier between the commune and the production team) or commune level. Within an overall system of central planning, communes were required to yield to the state a specified quantity of produce each year—either in the form of taxes or procurement quotas—as well as combining efforts with others in their given territory to achieve self-reliance through their own production. Even if one follows the path cut by Oi (1989, 4) in analyzing Chinese agriculture at the production team level, which she makes the case for because it is "the unit of organization where peasants live and work," the reorganization of agriculture that was initiated in 1978 was still significant.

Although there is disagreement in the literature about the locus of the impetus for downsizing agricultural production,[27] there is consensus about the extent to which it was carried out. As Nolan (1995) characterized the

26. Sato (2003, 187) sets forth a much higher figure here—17 percent for a specific community under study. This included both the land used by the private plots (8.6 percent) and the area set aside for growing feed for the households' livestock (8.3 percent). In discussions of this latter production in the Soviet case, this area was typically included in calculations of the collective area. Hence, if we exclude this feed-growing area from the calculation, this case study coincides generally with Riskin's (1987) figure.

27. Most of the literature describes this process as one which was effectively controlled by the party and state (e.g., Unger 2002; Nolan 1995; Fewsmith 1994, among others), yet a few scholars have set forth the thesis that it was entirely a product of the peasants' push for decollectivization (Kelliher 1992; Zhou 1996).

reform more generally, what began on an experimental basis in Anhui Province spread over the next few years to encompass virtually the entire country. The net result was that by 1983 approximately 98 percent of agriculture was farmed on a household basis (Unger 2002, 103). The organization of work had essentially been entirely decollectivized. Moreover, the production process was controlled by the household. Production procurement contracts were made between the production team and the household. These contracts effectively stipulated the land and other inputs to be supplied by the production team, as well as production quotas to be achieved by the household and the compensation they would receive (Hartford 1985). Within the parameters of a planned economy, production had been devolved to a small scale.

What distinguished the new land tenure system from a complete return to private property relations was that the village community retained ownership rights to the land. The objective here was to avoid a reconcentration of the land. The village community also controlled the terms on which the land was contracted out to households. Land was sometimes distributed on a per capita basis, and other times on the basis of a combination of household size and labor force participation. In addition, the village community received the "Ricardian rents" derived from the land. These rents were to be used for community needs (Nolan 1995, 191). Hence, although some aspects of what had heretofore represented socialist agriculture in China were transformed as a consequence of these measures, several key features of that model were retained.

However, this was only one of a series of changes that took place in the countryside in this period. As part of the effort to redress the problem of stagnating production levels, a new emphasis was placed on increasing the inputs available to peasant farmers. Farm input availability rose notably after the reform was begun: between 1979 and 1983, the average annual growth rate of chemical fertilizers was 13.6 percent, for farm machinery (horsepower) it was 9 percent, and for estimated total nutrients it was 6 percent (Riskin 1987, 297). Clearly, in the lead here were chemical fertilizers. Inadequate supply seems to have been the principal limitation on their usage prior to the mid-1970s. Fertilizer imports were a partial solution. More important, national fertilizer production doubled between 1970 and 1975, and between 1975 and 1979 it doubled again (Bramall 1993, 291). Thus, before the reform, a significant initiative was under way to attack the issue of production levels by increasing access to inputs.

Access to inputs was further expanded with the raising of farm produce prices. With better procurement prices, farmers had greater wherewithal to purchase the larger quantities of inputs that were now available. There were several components in the new pricing policy that together brought about an improvement in the terms of trade for agricultural producers. First, prices paid for goods purchased by the state through the procurement (or quota) system were augmented: in 1979, they were augmented by 22.1 percent over 1978 levels (Bramall 1993, 285). Although the level of increase varied somewhat by product, the general pattern was consistent. Prices for above-quota goods sold to the state also rose markedly: in the case of grain and oils, by approximately 20 percent (calculated from Bramall 1993, 285). These policies were complemented by a reduction in procurement quota levels. Finally, given that there was no corresponding rise in the prices charged by the state for industrial goods sold in rural areas, the net result was a notable improvement in the terms of trade for the farm sector.

This situation experienced a reversal in the mid-1980s. In the context of a deepening recession and serious disagreements within the Communist Chinese Party (CCP) concerning the future course of China's economic reform, government policy shifted after 1984 and quota procurement and farm gate prices dropped. The shift in policy had a real effect on income levels and regional disparities, both of which will be addressed later. Nonetheless, it is essential to note here that during the 1978–84 period improved terms of trade were a central piece of the government's overall proagriculture strategy.

Another piece of that strategy was, at least ostensibly, termination of the state-dictated policy of territorial self-reliance in food grains. This was meant to translate into a new emphasis on comparative advantage that would, in principle, allow farmers to profit from producing commodities particularly suited for their region and which might well fetch better prices than the grain that they had previously focused on to ensure that the goal of self-reliance was met. There were a number of problems with this policy modification, though, starting with the fundamental contradiction between the state's efforts to continue to plan the economy and this allowance for individual (or production team) choice about production priorities. This approach was bound to result in the substitution of less remunerative crops by more remunerative ones with the potential that held for drops in the production of essential foodstuffs. The government's move to freeze the level of interprovincial grain transfers, as well as that of grain resales to non-grain-producing farmers in early 1982 (Hartford 1985), provided evidence that this potential was already close to being realized. Yet the initial loosening of the previous production requirements, even if ultimately only partial in nature, formed a key component of the reform set in motion in 1978.

The last major change that was part of the reform policy package was the reopening of farmers' markets. Between 1949 and 1979, marketing of agricultural produce, especially food grains, was largely controlled by the state. For much of that period, produce sales outside of official channels were illegal. With the reform came an expansion of free markets in rural areas and the opening of private markets in towns and cities. The liberalization of produce marketing, combined with the loosening of local self-sufficiency requirements, meant that peasants could produce more lucrative crops and market them where the prices were highest. However, as Oi (1989, 212) points out, it also led to growing uncertainty for farmers, as they were no longer guaranteed minimum incomes, and they found the situation less than ideal.

What was the outcome of the monumental shifts in agricultural policy that were initiated in 1978? Probably the most significant outcome was the further diversification of production in rural areas. With the loosening of restrictions on production, as well as on each household's allocation of the labor of its members, animal husbandry took off, sideline production soared, agriculture grew, and product mix became more varied. Growth was greatest in sidelines (which operated at both the individual and collective level), averaging 18.6 percent a year between 1978 and 1984 (Riskin 1987, 291). In addition to the resultant increase in goods this made available, it also brought about a more efficient allocation of labor in rural areas as "surplus" farm labor became free to move into higher-paying areas of work. Yet agriculture, strictly speaking, also experienced a remarkable level of expansion: animal husbandry increased by an average of 9.4 percent annually during this period and agriculture alone by an average of 6.7 percent annually (Riskin 1987, 291).

China scholars debate the principal reasons for this growth. The main area of disagreement concerns the extent to which decollectivization led to the expansion of production.[28] But, there is little disagreement that it was dramatic and that it was positive in a number of senses.

28. Nolan (1988), Putterman (1988), and Selden (1988) number among those who argue that decollectivization played the key role in production expansions experienced after 1978. Bramall (1993, 2000), Unger (2002), and Riskin (1987) argue that the major expansion in growth began before decollectivization took off, instead pointing to improved producer prices, greater availability of inputs (see especially Bramall), or a general relaxation of state pressures on peasants (see especially Riskin).

Perhaps most important, it brought about an increase in peasant incomes.[29] Unger (2002, 112) notes that, for the country as a whole, Chinese statistics show "a rise of 67 percent in real per capita peasant incomes between 1978 and the end of 1982."[30] This assisted in reducing the urban/rural disparity in income generation that had existed for some time. Travers (1985, 112) states that already by 1981 discrepancies in consumer goods purchases between urban and rural areas had dropped to 7 to 1 from the 10 to 1 they had been in 1978. And, according to Riskin (1987, 293), "while average per capita 'disposable income' of urban workers and employees grew by 42.7 percent between 1978 and 1983, net per capita income of peasants was rising by 98.4 per cent." Thus, the shift in agrarian policy making had brought about a major improvement in the situation of the peasantry overall.

However, the downward modification of producer prices that was initiated in the mid-1980s curtailed the pattern of income advancement for farmers. Producer prices continued their downward trajectory during the 1990s, falling as much as 30 percent between 1997 and 2000 (Unger 2002, 173). Simultaneously, government procurement quotas rose. In addition, taxes imposed on agricultural producers were augmented, which hit those in poorer regions especially hard, as it was they who relied most heavily on agriculture to ensure their livelihoods (Lu and Wiemer 2005). Those in better-off regions were more likely to be able to bolster their economic situation with expanded sideline production. Moreover, the cost of agricultural inputs rose greatly, reversing the improvement in the terms of trade between industry and agriculture that had occurred between 1978 and 1984. Not surprisingly, rural per capita income stagnated after 1985 (Unger 2002), and the earnings gap between rural and urban areas widened once again.[31] But ongoing growth in off-farm income masked the decline in real agricultural income. This dynamic continued throughout the 1990s and into the new century. As

29. Zhong (2007) argues that, even more than agricultural growth, it was the increased diversification in production that led to improved peasant incomes.

30. Unger (2002) also notes that in many of the villages included in his survey a good part of this increase came before decollectivization, thus coinciding with the position of Bramall (1993, 2000) and Riskin (1987). Riskin (1987, 293) specifically states, in fact, that approximately one-fifth of the increase in real rural per capita income was due to price increases.

31. According to Fan and Chan-Kang (2007, 49), "in 1985 the average rural income was 60 percent of urban income, but the ratio dropped to 33 percent in 2002." And, Patel and Eisenburger (2003, 11) note that three-fourths of China's "impoverished," which were calculated at 100 million people in the late 1990s (as they earned less than $1 per day in purchasing power parity), were rural dwellers.

a result, despite the Chinese government's claims that rural poverty decreased during this period, Riskin and Shi (2001) argue that there was an increase in the absolute number of rural poor. They conclude this because the poverty rate remained virtually the same while the rural population grew, leaving ever more people in this category of the population.

At the same time, the economic reform affected China's rural regions differentially, especially after producer prices fell in the mid-1980s. While rural poverty dropped slightly in eastern China between the mid-1980s and the mid-1990s (from 9 to 5 percent), and more significantly in the central region (from 20 to 13 percent), it grew in the western region (from 26.5 to 31 percent) (Unger 2002, 171).[32] Likewise, the Gini index for rural China grew notably between 1978 and 1995.[33] Hence, it was peasants in those regions close to urban areas who had benefitted most from economic liberalization, as it was they who had been able to take advantage of growing crops, such as vegetables, that fetched higher prices in farmers' markets. In contrast, those far removed from urban markets, and particularly those in grain-growing areas, were hard hit. While this may simply represent the exaggeration of an historical tendency, economic polarization within villages also grew.

Over time additional problems associated with the economic reform, and especially with the decollectivization of agricultural production, became apparent. These included conflicts between central planning and market functioning, destruction of collective property (especially forests, land, and water-control systems), increased insecurity of peasant livelihoods, and that the parcelization of land made it difficult to employ machinery in agricultural production because of the small size of peasant plots. All of these issues raised questions about the sustainability and desirability of the reform model. Indeed, the longevity of this model was under debate by China's reformers. A group of them sought to push the process of change much further along in the direction of market liberalization (see Unger 2002; Fewsmith 1994). A key point of contention was whether landholdings should become private

32. And Fan (2006, 83) points out that "the southwest and northwest regions together accounted for about 70 percent of the rural poor in China in 1996—a 40 percent increase compared to 1988."

33. Fan and Chan-Kang (2007, 49; see also Chang 2006) state that the Gini coefficient for China as a whole grew between 1980 and 2000 from 0.33 to 0.46. They go on to state that while "both rural and urban inequality have increased, the inequality between rural households was much larger than the one between urban households." And, after a survey of recent studies of inequality, they conclude that "the rise in China's total inequality was mainly due to inequality between provinces . . ." with the eastern provinces being the more privileged among them.

property. Those who argued for privatization were heavily influenced by Western economists who saw the root of many of China's (and other non-capitalist countries') problems laying in the social ownership of property. Their opponents did not share this perception.

The playing out of this debate in the agricultural sector could be seen in the changing laws defining the length of time households had use of the parcels they were allotted. Initially, the contracts were short term. But in 1984, they were lengthened to fifteen years to discourage predatory use of the land and to encourage investment in it (Riskin 1987). In 1993, the duration of contracts was further lengthened to thirty years, clearly representing a step in the direction of private ownership. Yet implementation of this latter decree encountered resistance at the village level as farmers wanted the configuration of landholdings to be able to vary with alterations in their family size and age breakdown.[34] And, multiple surveys documented the strong preference of China's farmers for collective ownership of land, more generally, which would ensure that they could never become entirely landless (Unger 2002). Moreover, those reformers who promoted a less radical model argued that privatizing the land would do little to resolve the problems that had emerged since 1978, problems of increasing social differentiation within and between regions, growing rural poverty, and lack of care for "the rural commons."

This debate was related to a larger one that was also under way concerning China's entry into the World Trade Organization (WTO) in December 2001. Of concern was its effect on various parts of Chinese society. As might be expected, there was a fair amount of disagreement about its consequences for farmers.[35] There was a fair amount of consensus, though, about which kinds of farmers were likely to benefit and which were less likely to. In the former group were those who produced fruits and vegetables—coincidentally, those who had benefitted the most from China's post-1985 reform process. And, this would especially be so for those producers who were located within easy access of markets. Grain producers, particularly those located further from markets, were likely to experience a disadvantage. Once again, these were precisely the farmers who had been hardest hit by the reform since

34. Patel and Eisenburger (2003, 20) mention that in some areas contract duration was reduced to two to three years because of high population growth.

35. Several of those who argued that its impact would be positive were OECD (2005) and Huang and Rozelle (2003). Among those who were concerned that its impact would be negative were Patel and Eisenburger (2003), while Fan (2006) argued that the government needed to undertake various efforts to ensure that it would not exacerbate already existing inequalities.

1985. Hence, without the implementation of some protective measures, the likelihood was great that the growth of interregional and intrapeasant inequality would continue unabated.

A partial shift in government policy was initiated in 2004, reflecting an awareness among policy makers of the negative consequences post-1985 reform policies had for farmers and their production. This shift could be seen in the elimination of the agricultural tax that fell so heavily on the country's farmers. Although scheduled to be phased out over five years, its implementation was begun quickly in the country's most intensely agricultural provinces (see Lu and Wiemer 2005; Heerink, Kuiper, and Shi 2007). In addition, grain prices were raised in 2003 and 2004, and grain producers began to receive a new income subsidy and other supports from the state. According to Heerink et al. (2007, 147), the initial results of these policies included expansion in the area sown in grain and in its production, increased income for farmers, and a reduction in rural poverty. Yet it remains to be seen how the Chinese state will be able to keep these favorable policies (at least in terms of grain prices and subsidies) in place once its "grace period" for complying with the "conditionalities" of WTO membership lapses.

Regardless of the outcome of these several debates, the shifts in peasant livelihood that characterized the post-1978 period highlight how crucial government agricultural policy was in their determination.[36] Between 1978 and 1985, the government, through its array of policy initiatives, sought to stimulate agricultural production. The chief actor in this equation was to be the peasant household. Although much of rural industry remained collectivized, the small farmer was the centerpiece of agricultural rejuvenation. In adopting this strategy, the Chinese government envisioned the country's peasantry as being economic actors. It saw them as having the potential to react favorably to economic stimuli, increasing their production overall and, in particular, their marketed production, thereby contributing to their country's more general economic boom. Thus, as the country's leadership set out to reconfigure socialism, the peasantry was to play an important economic role in the new model of development being drawn up there.

For most of the post-1985 period, the peasantry no longer held the more favored position that had characterized the earlier years of the reform. It was also apparent that the government was increasingly willing to permit social

---

36. In suggesting the impact of China's entrance into the World Trade Organization on rural dwellers, Fan (2006) makes this point quite strongly.

differentiation to proceed apace in the countryside, something it had strived to ameliorate in the Maoist period. Furthermore, in recent years, the problem of land being sold out from under farmers by the township in which it was located to be used for urban development, without adequate compensation to the farmer and without his or her acquiescence, was added to that of privatization in raising concerns about land security.[37] New legislation was put in place in 2003 that was designed to protect farmers and their land from urbanization. Yet it was unclear how effective it would be in addressing this problem.[38] Then, in October 2008, the CCP announced a major shift in agrarian policy that would permit farmers to lease or to transfer land-use rights (Yardley 2008). Government-run land markets quickly began to open up. Whether this policy shift might lead to reduced inequalities between urban and rural areas as its proponents argued, or growing landlessness in the countryside as its critics feared, would take time to become apparent. So, too, would the viability of the peasantry as an economic subject over the medium to long term.

Conclusion

This brief comparison of the transition toward the market that Russia and China have pursued illustrates their widely varying trajectories. It reveals the effect of those distinct trajectories on diverse social groups within society—with that of small farmers being of the most interest here. Hence, it also serves as a basis for reflecting on the question that lies at the intersection of the several bodies of theorizing introduced earlier: What impact does the transition from orthodox socialism to permitting some degree of market relations to operate within a country have on the peasantry—or small agricultural producers?

As described above, Russia's post-1990 leadership was committed to embarking upon a rapid retreat from socialism. A central part of that effort

37. In fact, Chen (2006) includes this issue in a list of five major concerns in China's rural development.

38. Chen (2006), Lu and Wiemer (2005), and Yao (2007) were relatively optimistic that the 2003 legislation would resolve this problem. But Walker (2008) was much less sanguine than they were about the potential for that legislation to be effective in this effort. And, Yao (2007) pointed to one of its unintended consequences: the undercutting of the village government's revenues for financing its social welfare policies in a context in which the village is the only real source of "social protection" in rural areas.

involved the implementation of a SA of the economy, which was supposed to open it up fully to market relations. Even though the end result fell far short of the leadership's hopes, in terms of eliminating obstacles to the expansion of the market, it did resemble Szelenyi and Kostello's (1996) political economic characterization of countries with a "capitalist orientation." As they had predicted—and as Polanyi's (1944) analysis of liberalism in England suggested would be the case—inequality increased drastically, as did the percentage of the rural population living in conditions of poverty. Moreover, the situation of small-scale agricultural producers weakened notably following the shift in political economic orientation. The government's policy package undercut the economic viability of their production at every turn, at the same time as it took apart the social safety net that might have buffered them from the changed economic environment. Here, too, a parallel exists with what Polanyi (ibid.) found for the initial expansion of market relations more than a century earlier: that the outcome of this process was the peasantry's growing marginalization. Whether or not Russia's small farmers would eventually be entirely eliminated (à la Lenin 1957; Polanyi 1944; and de Janvry 1981), however, remained to be seen.

The CCP leadership, in contrast, opted to reconfigure socialism there following Mao's death. The course they steered initially allowed for local markets in the context of a redistributive economy. As expected by Szelenyi and Kostello (1996), this had the effect of raising peasant incomes nationwide. Thus, between 1978 and 1985, inequalities were actually reduced and the prospects of the peasantry, as an economic sector, were improved. Yet subsequent changes in policy moved China into a socialist mixed economy. With this shift came an increase in inequality between regions and even within peasant communities, as those close to major urban centers thrived and those in more remote regions experienced relative economic decline. Clearly, the Chinese state had, during the second of these two periods, decided to let certain parts of the peasantry benefit from increased market relations, while leaving others to languish in this new economic context.

These two cases place in bold relief the importance of the position adopted by the state with regard to the role the peasantry, or the small farm sector, will assume within the economy. As Russia and China initiated their transitions toward the market, their respective leaderships made conscious decisions about the prominence agriculture would have in their new political economies. Because of the weight that small farmers—in the form of cooperative members—had previously held in each country's agricultural sector, decisions were also made about whether they should be relied on to contribute to the economy's reactivation, as in China, or whether they should be left to fend for themselves while other sectors became responsible for generating economic dynamism, as in Russia.

Although the era and the context were quite distinct from those described by Polanyi (1944), Friedmann (1978), and Mann (1991), the effect of the state on the prospects of small-scale producers was, as they had noted, also significant in the recent period of political economic change in Russia and China. Polanyi's focus was not on small farmers per se. Nonetheless, their fate played a central role in his sociohistorical analysis. After all, their transformation from agricultural producers into an urban working class, which the British state was crucial in facilitating, was essential for the development of capitalism in England. In contrast, Friedmann (1978) and Mann (1991) were writing specifically about the uneven nature of the advance of capitalist relations of production in agriculture. They were speaking directly to the ongoing debate concerning the future of the peasantry, given the expansion of capitalism into agriculture. The spread of markets in Russia's and China's economies has also affected each country's small farmers. But those effects have differed dramatically, depending on whether a redistributive orientation remained hegemonic. In the case of Russia, it no longer was. Hence, agricultural producers, especially small farmers, became more marginalized. In the case of China, it continued to be hegemonic and the situation of small farmers was more varied. There, too, the incursion of the market had hurt some, while it helped others to prosper. And, state policies had been crucial in defining this variation.

It is now time to look more closely at the process of political economic change in Nicaragua and Cuba. In both countries, the overarching political economy (i.e., whether a redistributive or capitalist orientation was hegemonic) was decisive in determining the vision of agricultural development and of small farmers held by policy makers. However, it will become apparent that other dynamics suggested in the literature on the agrarian question also came into play in these two cases. And, having Russia and China as a backdrop will provide a frame of reference about the range of options in agricultural policy making pursued by those countries in transition to the market.

PART 2

Nicaragua's Rapid Retreat from Socialism

3

The Economic Strategy of the Post-1990 State

The shift in political economic orientation of the Nicaraguan state began with the inauguration of the Chamorro regime in April 1990. A modification in the balance of forces of the distinct social groupings within Nicaraguan society had, effectively, culminated in the electoral victory of the anti-Sandinista alliance, UNO, or Unión Nacional Opositora. That victory gave majority control over the state to representatives of the capitalist class, thereby opening up a new stage of the struggle for hegemony that had escalated during the Sandinista era. Thus, from a government that had initiated a transition toward socialism in Nicaragua under the Sandinistas, the baton passed to a government that was committed to bringing free-market economics back to the country.

That there was a shift under way in the reigning political economy was put in evidence in a variety of areas of government policy making. These ranged from the economic sphere (which will be analyzed at length) to the social/ideological sphere, as expressed, for example, in the transformation of educational policies. Most of the changes arising from the new political economic framework coincided closely with the policy requisites of the key governments, most especially the United States, and the international lending community that the Chamorro government was eager to establish relations with. Hence, the restructuring of the economy and society that the government embarked on in 1990 was part of its larger project of bringing about a rapid retreat from socialism, while also endearing it to those it sought to ally with and borrow money from.

The country's improved standing with the international financial community represented a stark contrast from the status quo of the 1980s. During a relatively brief honeymoon period (1979–81), the development plans of the Government of National Reconstruction—which was constituted following the Sandinistas' overthrow of the Somoza regime—were endorsed, albeit tepidly, even by the World Bank. After this time, however, Nicaragua's relationship with the international lending community became extremely restricted and continued to be so throughout the Sandinistas' tenure in power. In large part because of the pressure the U.S. government brought to bear on crucial multilateral lenders, by the mid-1980s, Nicaragua's government could only look to a limited number of bilateral sources for loans and assistance to fund its development program.[1]

That development program was to be based on a mixed economy, with a state and cooperative sector that were initially formed with properties confiscated from the Somoza family and its close allies, and a private sector that had producers of all sizes and organizational types. Part of the profits from state enterprises was to be distributed socially through the expansion of social services. The remainder was to be invested in infrastructural development geared toward agroindustrial enterprises that the government was building with international assistance. These new enterprises were designed to modify Nicaragua's role in the international economy from an exporter of raw materials to an exporter of processed agricultural goods. Yet export production would now have to share many productive resources it had previously monopolized with the sector of production geared to the domestic market. That is, the Sandinistas' program was oriented toward changing the nature of the country's productive structure, as well as redistributing productive resources and the surpluses that were generated by them.

Nonetheless, foreign assistance began to shrink with the escalation in U.S. hostilities as of 1981. And the defense effort consumed a growing percentage of the national budget as the Contra war advanced from primarily consisting of skirmishes to rendering uninhabitable (and unproductive) larger and larger parts of the central mountain region from 1981 to 1984.[2]

---

1. See further, Stahler-Sholk (1990), Conroy (1987), and Kornbluh (1987).

2. Remnants of Anastasio Somoza's National Guard, which the Sandinistas had swept from the country in 1979, actually initiated their "counterrevolutionary" activities as early as 1980. During the 1980s, the Contra forces, which were financed, armed, and organized by the U.S. govern-

Thus, the Sandinista government found itself forced to shift economic gears. Over the next several years, it implemented a series of economic reforms designed to bring about greater control over inflation and fiscal imbalances while maintaining intact the larger objectives of the development plan.

The first section of this chapter will provide a brief look at the changes that took place in Sandinista economic policy during the 1980s. Then I will describe the Chamorro government's economic reform, which will clarify the significance of the shift set in motion in 1990. The latter review, which is the principal focus of this chapter, will analyze in much greater detail the multiple components of the reform of the 1990s and their implications for the country's peasantry.

I will show how, over the course of the 1990s, small farmers—that is, the vast majority of Nicaragua's agricultural producers—were increasingly cut off from receipt of the resources that were crucial for their production and, therefore, their livelihoods. For many if not most of them, access to these resources had been one of the gains they had experienced under the Sandinistas in the 1980s. With the rapid retreat from socialism brought about by the Chamorro regime, and the peasantry's resulting loss of these, and other, resources, they became more marginalized. In essence, in the following pages, it will become clear that Polanyi's (1944) prediction about the dire social consequences for the peasantry of spreading market relations were largely borne out with the reopening of Nicaragua's economy to the international economy as of 1990.

The Sandinistas' Economic Reforms

Within a few years after the Sandinistas assumed power over a government and country that had been ransacked by the outgoing Somoza regime, the trend toward economic recovery was evident. Yet that process was cut short by imbalances emerging in various areas of the economy. In addition to the

---

ment, increasingly drew support from parts of the landowning class and its most loyal workers who opposed the agrarian reform and other aspects of Sandinista policy. In some regions, the Contras also drew support from parts of the large and middle peasantry, who feared a broadening agrarian reform and had been hurt by the Sandinista government's grain commercialization policy and the military draft. In the war zones especially, some sectors of the poor peasantry who had also been negatively affected by the grain policy and the draft supported them as well. For more information on the Contra war and its relationship with the rural population, see Horton (1998), Spalding (1994), Escuela de Sociología-UCA (1987), and Robinson and Norsworthy (1987).

economic and military aggression orchestrated by the United States, problems associated with economic planning contributed to the expanding budget deficit and inflation (see Stahler-Sholk 1990; Pizarro 1987). These problems included having placed too much faith and financial resources in development projects that required excessively long periods of time to come on line and major sacrifices in other sectors of the economy to keep investment funds flowing to them once foreign assistance dropped off, as well as mistakes made in setting producer prices for domestically grown basic grains, which undercut production incentives. By late 1984, it was obvious that serious changes would have to be made if control over the economy was to be secured once again. Thus, in 1985, the Sandinistas implemented the first of several economic realignments. The initial steps toward reform entailed phasing out price subsidies, establishing a system of regular salary increases, devaluing the exchange rate, augmenting import duties on nonessential items, liberalizing basic grain prices, incentivizing agro-export production, lowering credit coverage from 100 percent to 80 percent of working capital, and reducing the fiscal deficit from 23 percent of the gross domestic product (GDP) in 1985 to 17 percent in 1987 (Stahler-Sholk 1990, 172).

When the 1985 policy adjustment did not succeed in bringing inflation under control but opened the way for a dramatic escalation of it—or in redistributing income in favor of salaried workers, in stimulating a serious increase in export production, or in containing other internal and external macroeconomic imbalances—it was followed up with a more comprehensive economic adjustment in early 1988. This adjustment had as its principal components a new schedule of controlled prices and salaries, a monetary reform, and a large devaluation and unification of the official exchange rate. As noted by Stahler-Sholk (1990, 198), the 1988 economic reform had a number of objectives: "The short-term objectives were to withdraw excess money from circulation, dampen inflationary expectations, and neutralize the monetary holdings of speculators and contras. In the longer term, the reform had five main goals: 1) realign relative prices; 2) reactivate exports; 3) decrease inflation by reducing public spending and the fiscal deficit; 4) induce greater efficiency in the formal sector of the economy; and 5) restore the normal economic function of salaries by increasing purchasing power."

The February 1988 reform had only limited success. This led to the implementation of further measures later in 1988 and 1989. These additional adjustment efforts moved Sandinista economic policy closer to more orthodox reform programs and were geared toward an advanced compliance with what were perceived to be the requirements of the European Union, as well as several other potential sources of financial assistance.

Although these reforms bore some resemblance to the structural adjustment (SA) measures adopted elsewhere in Latin America during this time, and to those that were subsequently implemented in Nicaragua once the Chamorro government came into office, they also had some notable differences. Included among the differences were a continuing effort to recover and sustain the purchasing power of the working class, to attain a balance between domestically oriented production and that destined for export, and to maintain (if no longer expand) the system of free health care and education that had grown massively from 1979 to 1983, and the fact that the reforms did not entail privatization of properties that had been nationalized or redistributed to the rural and urban poor but had been, instead, accompanied by a deepening of the agrarian reform.[3] Also in strong contrast to SA programs imposed elsewhere and under the subsequent regime in Nicaragua, these reforms were implemented without the benefit of a serious inflow of foreign capital to underwrite them. Clearly, they were designed to restore financial equilibrium to the economy, while retaining key elements of the social program that had been at the core of the Sandinista government's development project.

However, following the electoral loss of the Sandinistas in February 1990, the new government that came to power had few compunctions about dismantling the social program that had been in place since 1979. Thus, the aspects of economic policy that had distinguished the Sandinistas' efforts at reform were laid by the wayside as a traditional-style SA was fully embraced.

The Chamorro Government's Orthodox Economic Reform

The economic adjustment program implemented by the Chamorro government in 1990 brought Nicaragua into the fold of those countries—then representing almost the entire Latin American region—that had embarked on the path of reforms that flowed out of neoliberal economic thought. By the early to mid-1980s, SA had become the sine quo non throughout Latin America for gaining legitimacy in the international economic community.

Even though SA was first imposed in the Latin American context after

---

3. This was especially true of the 1985 adjustment measures.

the 1973 coup in Chile, it was not until the 1980s that it became the predominant economic model throughout the region. By that time, bilateral and multilateral lenders had succeeded in bringing most of the area's leaders around to their perspective of the way Latin America's economies should be run. How had this occurred? What did the policy shift consist of? And, what did it look like in the Nicaraguan context?

The terrain for SA to gain the day was prepared by the economic crises that characterized most of the region's economies by the early to mid-1980s.[4] After several decades of impressive economic growth throughout the region, the boom in oil prices in 1973 triggered an economic shortfall in the area's oil-importing countries. Given an influx of funds from the oil-producing countries, the multilateral lending agencies and banks stepped in with loans to resolve the problems caused by the new, massive oil bills. Yet, by the early 1980s, it was clear that many Latin American countries could not repay their foreign debts, which had mushroomed since 1973.

A variety of factors had contributed to their debt crisis. Among them were shortcomings in the import substitution industrialization (ISI) model. What had begun as an attempt to reduce these countries' dependence on the international market (which was experienced so poignantly during the Great Depression and World War II when primary goods prices fell, wreaking havoc throughout their economies, and imports became ever more difficult to obtain) by initiating the manufacture of industrial goods locally, had resulted in the formation of an industrial sector for nondurable consumer goods and large import bills for the capital goods that were needed for the former's production. At the same time, the export of primary goods had remained important, so that inputs for industry could be imported. Nonetheless, the terms of trade turned against primary products as all of these other economic dynamics were taking hold, further weakening these countries' ability to keep industrializing and to stay abreast of their now enormous foreign debts.

In the face of the threat of Latin America's failure to pay its debt, the United States and the multilateral lending community offered each of the debtors the possibility of debt restructuring and the provision of new funds. But, these were contingent on the acceptance of major changes in their economic policies, in the form of stabilization and SA programs. In some cases,

4. For a variety of perspectives on SA, see Williamson (1990), Loxley (1986), Haggard and Kaufman (1992), and Veltmeyer, Petras, and Mieux (1997).

these changes were brought about by recently installed military regimes, as in Chile and Uruguay in the 1970s. In others, democratically elected governments opted for this course instead of the more radical alternative of refusing to pay their debt—with all of the likely consequences that would entail—as in Mexico, Costa Rica, and Ecuador in the 1980s. Despite these regime-type differences, and the variations between the specific programs designed in each country, they bore a number of features in common: the imposition of government austerity measures, trade liberalization, currency devaluation, and privatization (Conroy, Murray, and Rosset 1996). Their stated objective was to foster economic recovery, while their implicit one was to ensure that these nations paid back their debts. Attainment of this latter objective would be facilitated by increased exports.

A number of assumptions underlay the central features of the economic model at the heart of SA. Among these was that economies functioned best where the state played a minimal role in them, echoing Adam Smith's theses of the eighteenth century,[5] which were core building blocks for classical economics. Closely associated was the assumption that countries ought to allow their historic "comparative advantage" to determine their production emphases, which had its origins in classical and neoclassical economics. That is, rather than nation-states ascertaining what their populations needed and seeking to produce as much of that locally as possible, it made more economic sense to promote the production of those goods that the country was "particularly suited for" and exchange these in the international economy for the goods their populations consumed but which the country was not endowed to produce. It made no difference if these "advantages" were historical artifacts, which were then posited as "natural" to protect the economic interests of the industrializing "north," as some critics contended (de Janvry 1981; Emmanuel 1972). Nor did it matter whether the international economy did not compensate many countries according to the labor invested in the goods they had been deemed "suited" to produce (Economic Commission for Latin America [ECLA] 1950). The point was to do away with the inefficiencies inherent in trying to produce all that was needed locally. And that meant eliminating the state-sponsored efforts to build and to protect new industries where they had not emerged "on their own" and to lift any barriers to trade that disadvantaged the flow of goods and capital that had prevailed before the era of state interventionism and protectionism (particularly as em-

5. See Smith (1946).

bodied in the ISI model). Three of the features of SA mentioned earlier—(1) trade liberalization, (2) currency devaluations, and (3) privatization—were specifically oriented to work toward these objectives. The fourth, the application of austerity measures, was supposed to reduce an even broader array of "unnecessary" government expenditures to allow for a less wasteful use of state resources, incidentally enabling the redirecting of more of them toward paying off the foreign debt.

In spite of the distinct circumstances leading up to the imposition of SA in Nicaragua, the Chamorro government's economic reform contained all of the features common to more orthodox programs.[6] I will analyze each of these in turn. But, rather than describing Nicaragua's entire SA program, what follows is an examination of the components of these features that directly affected agriculture, and particularly small producers, there.[7]

Before doing this, however, it is important to note that though the new government was clear in its commitment to carry out SA, it did not have a similarly clear vision of rural development and, especially, of the role of the small farmer within it. It sought to increase agro-exports and to do whatever was necessary to pursue what it considered Nicaragua's comparative advantage to be. But those efforts did not form part of a larger model of rural development, much less one that specified the contribution of each sector of the rural population to it. Instead, its vision of small farmers was implicit in its SA program: to the extent that they could contribute to agro-export production they were seen in a positive light, despite not being provided with the resources needed to do so. Where they could not or did not, they were to be addressed—if they were addressed at all—as a social category that needed the limited handouts that were to be given to "marginal" groups who were not among the "new economic actors" designated to revitalize the economy and who would, therefore, have to bear the brunt of SA. These handouts were built into the international loans and foreign aid that financed the SA.[8] I will now discuss concretely the ways in which small farmers were made marginal through the process of SA.

6. In addition to the measures enumerated in Chapter 3, it was similar to other SA programs in obtaining substantial foreign assistance from multilateral lending sources to fund its implementation and to assist the government in fulfilling its multiple requisites (cf. Renzi 1996; Acevedo Vogl 1993).

7. For a selection of studies of SA in Nicaragua, see Acevedo Vogl (1993), Evans (1995), Saldomando (1992), Neira Cuadra (1996), and Stahler-Sholk (1997).

8. Spalding (2007) describes the evolution of these aid programs in the post-1990 period and points to the extent to which they were linked to, and in some cases required for receipt of, international loans.

The Application of Austerity Measures

Austerity measures typically play a central role in SA programs, as government budget deficits are one of their key targets, given the role of these deficits in generating inflation and fiscal imbalances. The Chamorro government's austerity measures affected various parts of its budget. Among other things, they resulted in major cuts in expenditures for social services. As a consequence, new charges were applied for formerly free medical attention and education at the same time as the actual services received were more limited. Another outcome of the measures was the massive layoff of government employees. Approximately a quarter of all state employees were let go during the first few years of the SA program (Renzi 1996; Evans 1995; Stahler-Sholk 1997).[9] All of these changes had an effect on the rural sector.

The austerity measures also brought about a significant drop in credit destined for agricultural production. For example, in the 1995/96 agricultural cycle, the crop acreage covered by credit from the National Development Bank (BANADES)—the state bank responsible for loans to the agricultural sector—was only 24.7 percent of what it had been in 1990/91 (see Table 3.1). This precipitous decline continued over the next few years so that the area financed in 1997/98 was only 9.3 percent of that of 1990/91. Yet this drop-off in state-sponsored credit was not implemented across the board in agriculture. True to the objectives of SA, export production was prioritized. While financing for coffee acreage (Nicaragua's most important export crop) in 1995/96 was 41.7 percent of what it had been in 1990/91, financing for bean and corn acreage (the country's two key food crops) in 1995/96 was 4.9 and 5.4 percent, respectively, of what it had been in 1990/91 (see Table 3.1). Partially as a result of this shift in funding priorities for different types of crops, producers did not bear the burden of the cuts equally. While small- and medium-sized producers had received 56 percent of agricultural credit in 1990, they only received 29 percent in 1992 (Acevedo Vogl 1993, 112). In contrast, the share of credit provided to large producers, whose production emphasis was on export crops, grew from 31 to 71 percent during this same period.[10] However, new collateral requirements, often including having for-

9. And, Spalding (2007, 7) cites a dropoff of 68.8 percent in public employment (including the armed forces) by 1999.

10. Rodríguez Alas (2002, 103) notes that the largest producer strata (of a five-strata breakdown) monopolized 71 percent of all rural credit (state and otherwise) in 1998, thus underlining the concentration of financial resources in this sector.

Table 3.1 Area financed by BANADES: Total and selected crops, 1990/91–1997/98 (thousands of *manzanas*)

	1990/91	1991/92	1992/93	1993/94	1994/95	1995/96	1996/97	1997/98
Export crops								
Coffee	80.3	66.4	38.4	37.8	43.5	33.5	22.7	23.7
Sesame	51.3	23.8	23.0	7.6	19.1	22.0	22.0	0.2
Food crops								
Beans	57.0	39.9	5.6	8.8	10.0	2.8	0.2	1.3
Corn	150.2	131.6	20.2	25.2	26.2	8.1	0.5	9.8
Total area	560.5	464.6	232.5	228.9	214.8	138.5	86.1	51.9

SOURCE: BCN 1998, 28.
NOTE: A *manzana* is approximately 0.7 hectares.

mal title to one's property, also undoubtedly had an effect on this latter tendency—as small- and medium-sized producers were at a distinct disadvantage in this regard.

Government austerity measures also translated into a general cutback in BANADES operations. This made it ever more difficult for small farmers to gain access to its services. Between 1989 and 1993, over 60 percent of BANADES branch offices were closed (see Jonakin and Enríquez 1999, 9). Further cuts were made between 1994 and 1997. Finally, in early 1998, BANADES closed its doors completely, leaving small farmers entirely excluded from state-sponsored credit.

State development banks have traditionally been the principal, if not the only, formal source of credit for small farmers throughout Latin America. Commercial banks have typically steered clear of small farmers, claiming an inability to absorb the cost of overseeing many small loans for which processing costs are the same as for larger loans. Given the resulting exclusion of small farmers from the formal loan system, and the potential political threat their marginalization from all of the key agricultural resources contained within it, a number of Latin American governments opened small rural credit programs aimed at this sector of the population in the 1960s and 1970s.

The Nicaraguan government was one of these. In response to the call for agrarian reform by the opposition Conservative Party, the Somoza government established its Rural Credit Program in 1959. During the 1960s and 1970s, however, its reach remained limited (see Enríquez and Spalding 1987, table 2). Under the Sandinistas, the program experienced a dramatic expansion, from the 28,000 families it had provided credit for in 1978, to more than 100,000 families in 1980, before settling at around 80,000 families in the first half of the 1980s. With the cutbacks in BANADES funding as of 1990, its small- and medium-scale clientele dropped to 10,815 (Jonakin and Enríquez 1999, table 3).[11] And its departure from the banking scene in 1998 reduced that number to zero.

Nicaragua's banking system had been nationalized shortly after Somoza's overthrow in 1979 and remained under state control until the change in government in 1990. In the 1990s, though, a private sector in banking blossomed. Nonetheless, as was true for Latin America in the pre-1980 period,

---

11. The public sector did offer credit to small- and medium-sized farmers through its Rural Development Program in the early 1990s. Yet, the population served by this program and BANADES combined still only reached in the neighborhood of that served by Somoza's program (Jonakin and Enríquez 1999).

the new commercial banks offered virtually no loans to small- and medium-sized farmers. The latter's only real alternative was the nontraditional financial sector that emerged in the 1990s. At the core of that sector were nongovernmental organization (NGO) programs that sought to reach out to those farmers who had been cut off from state-sponsored credit. But their resources were limited, thus the number of farmers included in their programs was as well, as was the coverage they provided to those who they did include (Jonakin and Enríquez 1999).[12] In sum, with the eclipsing of BANADES in the 1990s, small farmers' access to agricultural credit was, likewise, largely eliminated.

Another agricultural resource that was essentially eliminated with the austerity measures that formed part of the Chamorro government's SA was technical assistance. This service can be helpful for improving yields of crops with which a farmer already has experience, but it is especially beneficial when new crops are introduced. Before the 1980s, this resource had only been available to those who could pay for it in Nicaragua. Economics determines access to technical assistance in most countries of the world. One logical consequence of this is that its employment is commonly concentrated among better-off farmers (see Thiesenhusen 1995; Kay 1994).

Under the Sandinista government, technical assistance was considered part of agrarian reform and agricultural development efforts. As such, its availability—free of charge—expanded dramatically. However, with the implementation of SA in 1990, technical assistance reverted to being a resource whose use was limited primarily to large producers who could pay for it. The only exception here, as well, was the technical assistance provided to the pool of beneficiaries of the NGO-sponsored rural development programs. But, as was true for those provided credit through these programs, they were only able to provide technical assistance to a relatively small number of producers because of their limited resource base.[13]

---

12. See also Davis and Stampini (2002), Boucher, Barham, and Carter (2005), and Baumeister and Fernández (n.d.), with regard to the extremely reduced access of farmers to credit.

Kay (2002) speaks to the growth of NGOs in postauthoritarian Chile, and their efforts to reach out to the peasantry who had been excluded from government programs in the period after 1973. But he also found that their limited resources restricted their reach within that marginalized population.

13. Davis and Stampini (2002) and Baumeister and Fernández (n.d.) provide data that illustrate just how limited access to technical assistance had become in the 1990s.

Trade Liberalization

A second key feature of standard SA programs, trade liberalization, was also implemented in Nicaragua. The rationale behind this measure is ostensibly to force greater efficiency among producers who have been protected by tariff barriers until now and to eliminate the burden on the state budget of these heretofore unprofitable enterprises. In general, this measure has been particularly oriented toward bringing an end to the existence of industries that were deemed only sustainable with such protective measures in place. But as Polanyi (1944) had found for the earlier period of liberalism, other economic sectors—including agriculture—were also subjected to the effects of the lifting of trade barriers.

Trade liberalization translated into the opening up of borders for increased exchange, with an eye to promoting products that could—through this test of sorts—demonstrate a clear comparative advantage. Because Nicaragua's historic "comparative advantage" was concentrated in primary products, especially agricultural ones, it was in this sector that production was expected to grow and exports expand. Given deteriorating terms of trade for some of Nicaragua's traditional export crops, most especially cotton, new agricultural products (nontraditional agro-exports, or NTAEs) were to fulfill this function.

The first step in liberalizing trade was the reduction of tariff protections. According to Stahler-Sholk (1997, 90), "the average nominal tariff protection was slashed from 43 per cent in 1990 to 15 percent in 1992."[14] Tariff protections generally hovered around this latter rate throughout the mid-1990s. However, with the signing of a series of free-trade agreements—with the other Central American republics, Chile, and Mexico—and pressure from international lending institutions, they then dropped still farther so that by 1999 they were 10 percent for finished goods from outside the region and tariffs were completely eliminated for intermediate and capital goods from the rest of Central America (BCN 1999, 127).

The elimination, or massive reduction, of tariffs typically opens the way for the importation of goods that had formerly been produced locally. In the case of basic grains in Nicaragua, drastically reduced tariffs were complemented by an overvalued currency. The result was to convert the remaining tariff levels for corn, beans, and sorghum into negative figures (Acevedo Vogl

14. See also Acevedo Vogl (1998) on this topic.

1998, 160–61). Because basic grains were the principal farm products grown and sold by small- and medium-sized producers, these economic trends pointed to a significant weakening of their position in the local market.

At the same time that tariff levels were falling, Nicaragua's adjustment program effectively eliminated the state monopoly on grain trading. That monopoly had operated through the National Enterprise for Basic Foodstuffs (ENABAS). ENABAS's functioning in the 1980s was known for its inefficiencies and for its flawed pricing policies (cf. Utting 1987, 1991; Spoor 1995). But, with its elimination, medium and small farmers were forced to reorient the marketing of their produce "from a sole marketing agent with guaranteed and fixed prices to . . . a newly established oligopolistic trading structure with a completely non-transparent pricing structure" (Eberlin 2000, 50).[15]

This, combined with grain imports—both in the form of donations and purchases (especially from elsewhere in Central America, given the new tariff-free status of exchange within Central America)—led to falling producer prices for domestically oriented production (López and Spoor 1993). The pattern of grain imports varied over the 1990s, beginning with a massive rise in imports in 1990; but the most sustained growth in imports occurred in the second half of the decade (see CEPAL 2001, 156, 158; and unpublished data, Ministerio de Agricultura–Forestal 2004). Increasing the country's reliance on grain imports was part of the comparative advantage logic. That is, it was assumed that the country's agricultural production should be geared first for export. If that meant importing food because it was perceived to make economic sense, then that was the path to follow. If following that path had a negative effect on prices for domestically produced grains, so be it.

Despite all of these disincentives, somewhat amazingly, production of basic grains expanded during the first half of the 1990s. As can be seen in Tables 3.2.1 and 3.2.2, although there was some variation from year to year in acreage dedicated to corn, and in production levels of this crop, its overall tendency was to grow. In contrast, bean acreage initially dropped off notably before undergoing an expansion. Yield levels pulled up bean production somewhat, though, so that overall production also rose. These various trends continued through the remainder of the decade.

15. See also Acevedo Vogl (1998).

The growth of corn and bean production was largely due to an increase in the population of producers resulting from the demobilization of Contra fighters and former members of the Sandinista military and their incorporation into agricultural production (Spoor 1994). Approximately 359,000 *manzanas* had already been turned over to ex-combatants from both sides by 1992 (de Groot 1994, table 3). This was equivalent to 12.8 percent of the land that was distributed to cooperatives and individuals through the Sandinista agrarian reform between 1979 and 1990 or to 49.2 percent of the area harvested with the country's major crops in 1992/93.[16] That the Contra war was brought to an end with the change in government in 1990, also, undoubtedly, opened up whole swaths of the countryside that had been uncultivable during much of the 1980s.[17] These swaths included some of the country's former key grain-growing areas. Peace and the terms under which it was achieved—that is, with a commitment on the part of the Chamorro government to provide land to ex-combatants—led to expanded production, even within the context of adverse economic policies. Nonetheless, those policies, combined with the rent-seeking behavior of the middlemen and small traders on whom farmers depended for the commercialization of their produce, as well as bad infrastructure—all of which were likely to be prevalent into the foreseeable future—continued to keep prices unfavorable for medium- and small-sized farmers (Eberlin 2000).

Trade liberalization, in its various forms, had the goal of promoting export production. Yet it was not until 1994 that the value of exports surpassed (and then only slightly) that of 1990 and 1995 before it took a notable step forward (see Table 3.3). Moreover, that step forward was principally due to the freeze that affected Brazil's coffee crop and pushed prices for this commodity upward (Stahler-Sholk 1997). With regard to traditional export products, the increased export earnings attained in 1995 held relatively steady through the late 1990s. Even with this improved situation, their annual value typically failed to reach the average amount of foreign exchange earnings

16. The first of these figures was calculated from de Groot (1994, tables 1, 3); the second was calculated from de Groot (1994, table 3) and MAG-FOR (1998, table 7). It should be pointed out, however, that the latter figure is not meant to suggest that the ex-combatants were responsible for cultivating this percentage of national acreage. Both figures were merely meant to give the reader a sense of what significance the amount of land given to the ex-combatants had.

17. Data presented in OIM et al. (1999) document the dramatic increase in acreage under cultivation that occurred between 1989 and 1995, bringing that figure back in the vicinity of where it had been in the late 1970s.

Table 3.2.1   Harvested area for Nicaragua's principal crops, 1990–2005 (thousands of *manzanas*)

Product	1990/91	1991/92	1992/93	1993/94	1994/95	1995/96	1996/97
Export							
Coffee	106.0	106.5	107.1	105.0	105.0	120.2	120.7
Sugarcane	60.5	60.0	56.0	54.1	59.7	64.0	71.4
Cotton	64.1	50.9	3.3	3.6	2.1	12.2	5.2
Sesame	50.6	23.8	27.0	27.0	39.1	52.8	37.4
Food crops							
Corn	250.0	282.2	250.0	312.8	280.0	320.0	398.5
Beans	150.0	135.7	130.0	164.7	172.0	150.0	171.3
Sorghum	36.0	37.9	47.0	49.4	45.0	19.5	51.4
Rice	54.5	55.0	63.0	81.4	83.4	89.9	96.6
Total acreage in key crops	771.7	752.0	683.4	798.0	786.3	828.6	952.5

SOURCES: BCN 2003, tables I.11–I.13; BCN 2007, tables 1–77, 1–12.

these goods had generated during the period between 1979 and 1984—a period of heavy state intervention in foreign commerce (see Table 3.3; and CEPAL 1997b, 364–65).[18]

Among the factors affecting the generation of foreign exchange earnings from traditional exports during this period was the volatility of prices in the international market for a few of Nicaragua's key crops. For example, the international price for cotton fell drastically in 1992, leading to a massive reduction in area planted after that time (see CEPAL 1996, table 55; and BCN 2000, table VI-8). Coffee prices, likewise, experienced a dramatic downturn in the early 1990s, a situation that was repeated at the end of that decade following the temporary "boom" of the mid-1990s. In the case of coffee, however, given its perennial nature, there was no corresponding downturn in crop acreage. Being subjected to this kind of price fluctuation is part of participation in the international economy, but this truism is much more accentuated for agricultural products.

As was true elsewhere in Central America, nontraditional products were to play a major role in the country's post-1990 export sector. Exports of such products increased in the early 1990s. However, here too, it was not until 1993 that the value of nontraditional exports (NTEs) grew substantially, after shrinking in 1991 and 1992 (see Table 3.3). Contrary to all expectations, the greatest contribution to export earnings was made in the manufacturing sec-

18. And, according to Acevedo Vogl (ca. 2004, 5), the value of agricultural exports as a percentage of total exports dropped from 80.3 percent in 1990 to 61.3 percent in 2001.

1997/98	1998/99	1999/2000	2000/2001	2001/2	2002/3	2003/4	2004/5
127.0	133.5	143.4	154.7	154.7	155.0	165.2	165.2
74.6	76.4	79.8	74.7	57.5	58.5	58.7	65.0
2.5	—	—	0.5	—	—	—	—
17.2	11.2	13.5	16.7	16.4	4.5	9.2	25.7
333.0	360.9	361.4	464.9	454.8	526.0	502.1	481.5
192.9	270.5	295.6	318.0	330.0	351.8	409.4	343.3
52.8	43.7	25.2	27.3	31.4	33.5	36.6	25.8
105.2	119.9	87.9	113.2	78.5	84.3	115.8	74.7
905.2	1,016.1	1,006.8	1,169.8	1,123.3	1,213.6	1,297.0	1,181.2

tor, rather than in the agricultural sector.[19] For example, agricultural products only represented 25 percent of NTE earnings in 1995, in contrast with the 67 percent represented by manufacturing (calculated from Table 3.3).[20] Even in 1997, a good year for agricultural NTEs, they still only generated 32 percent of NTE earnings compared with the 63 percent generated by manufactured goods.[21]

Furthermore, as was true elsewhere in the region, despite their expansion, agricultural NTEs still only generated a small proportion of total export earnings: approximately 7.1 percent in 1995 (calculated from Table 3.3).[22] Manufacturing generated approximately 19.2 percent of export earnings that year. But, most important for this study, with credit and technical assistance essentially unavailable to small producers, they were largely cut out of any growth that might take place in the NTAE sector. Hence, one more potential avenue to economic viability was closed off to them.

19. The export of manufactured goods did, nonetheless, experience a contraction after 1997, which continued through the end of the decade. It was not until 2003 that manufactured NTEs surpassed their 1997 levels. Then they grew steadily over the next few years.

20. And this was using the Chamorro government's definition of NTEs, which included tobacco and livestock. Arguably, these were not new to Nicaragua's export repertoire.

21. This figure averaged 37.3 percent in the first half of the 2000s (calculated from Table 3.3). Hence, although the presence of agricultural goods among NTEs had grown, manufactured goods still remained significantly more important.

22. By the turn of the century, this figure had grown to 14.1 percent, and by 2005, 17 percent. Nonetheless, even with this growth, their importance in overall export earnings remained relatively limited. For comparable figures elsewhere in Latin America, see Barham et al. (1992).

Table 3.2.2   Production levels of Nicaragua's principal agricultural goods, 1990–2005 (thousands of *quintales* [QQ])

Product	1990/91	1991/92	1992/93	1993/94	1994/95	1995/96	1996/97
Export							
Coffee	601.0	1,033.1	721.2	920.0	920.0	1,200.9	1,099.7
Sugarcane[a]	2,794.2	2,525.7	2,219.0	2,468.0	2,852.7	3,517.9	4,014.9
Cotton[b]	1,775.6	1,462.3	78.5	88.8	65.8	431.3	153.7
Sesame	281.5	185.6	170.0	216.0	375.4	417.9	222.8
Food crops							
Corn	4,375.0	5,079.6	5,000.0	6,256.0	5,320.0	6,400.0	7,103.3
Beans	1,200.0	1,275.6	1,235.0	1,688.8	1,840.4	1,500.0	1,647.2
Sorghum	1,152.0	1,421.3	1,579.2	1,729.0	1,575.0	777.4	2,058.1
Milk[c]	7,837.5	6,313.2	7,838.0	8,731.8	9,024.2	13,027.7	12,820.5

SOURCES: BCN 2003, tables I-11–I-14; BCN 2007, tables 1–11, 1–12, 1–13.
NOTE: 1QQ = a hundredweight.
[a]Cubic tons.
[b]Unprocessed cotton.
[c]Thousands of gallons. This only refers to the milk that was brought to milk storage centers. In the early 1990s, this represented approximately 17 percent of the milk actually produced (calculated from MAG-FOR 1999, table 12). The percentage of milk to reach storage centers increased over the 1990s, as milk exports increased. Moreover, the entry of a multinational corporation (Parmalat) into the Nicaraguan dairy market in 1999—which came to purchase and process a noteworthy part of national milk production—would have accelerated this process.

Currency Devaluations

Currency devaluations also formed part of the orthodox SA package. The devaluations were supposed to stimulate an increased demand for exports and a reduced demand for imports, as well as to attract foreign investment by reducing the dollar value of wages. In theory, all of this combined would improve a country's balance of payments.

The Chamorro government's SA program contained a series of currency devaluations. In addition to promoting export production, they were designed to force producers to be more efficient in their use of now more costly imported inputs. To ameliorate the effects of the devaluations on export production, though, the potentially negative consequences of more expensive imported inputs were mitigated by a new export promotion law. The law gave import duty and sales tax exemptions to exporters, as well as income tax exemptions to exporters of nontraditional commodities.

Domestically oriented crops and their producers did not receive similar protections. So their production costs rose, while prices for their goods remained stagnant or fell (Acevedo Vogl 1998). For example, FUNDENIC-INDES (1997, 102) refers to one period (January 1993–October 1994) during

1997/98	1998/99	1999/2000	2000/2001	2001/2	2002/3	2003/4	2004/5
1,433.7	1,439.3	2,083.3	1,808.5	1,271.7	1,152.3	1,795.8	1,100.6
4,125.9	3,805.1	4,055.8	3,877.0	3,459.6	3,431.2	4,565.6	4,429.7
58.2	—	—	7.7	—	—	—	—
147.5	69.2	103.1	114.7	153.2	34.8	81.3	225.1
5,809.5	6,610.3	6,397.2	9,052.7	9,241.6	10,886.8	12,048.8	10,030.2
1,573.6	3,279.7	2,958.9	3,800.9	3,904.5	4,289.3	5,044.1	3,982.7
1,410.7	883.9	947.7	988.7	1,275.1	1,362.5	1,382.1	1,133.9
12,300.4	10,494.9	10,415.4	10,919.5	14,601.8	16,467.4	20,432.0	20,704.2

which the producer price for corn rose by 29 percent, while the price for two of its key inputs, fertilizer and Malathion, rose by 320 percent and 177 percent, respectively. Thus, exclusion from the selective exemptions that Chamorro's government offered, represented further discrimination against basic grains producers.

Privatization

Finally, privatization of state-owned industries and farms was also central to orthodox SA programs. It was supposed to eliminate what were perceived as unprofitable state enterprises, opening up new opportunities to the private sector at the same time as reducing the state's budget deficit.

Nicaragua's privatization process was initiated with the creation of a state agency, the National Corporation of the State Sector, which had as its principal task to dismantle the 351 existing state enterprises (Renzi 1996, 19). Of these enterprises, 22.8 percent pertained to the agricultural and livestock sector of the economy. As the process got under way, popular pressure forced the inclusion of state farmworkers into the pool of recipients. They were to receive approximately 25 percent of the stock of each of the privatized enter-

Table 3.3   Value of Nicaragua's principal exports, 1990–2005 (millions of U.S. dollars)

	1990	1991	1992	1993	1994	1995	1996
Traditional goods	261.7	211.8	171.5	178.5	218.1	332.2	329.7
Agricultural	238.9	188.7	150.4	124.4	171.7	247.3	247.7
Others	22.8	23.1	21.1	54.1	46.4	84.9	82.0
Nontraditional goods	68.8	60.6	51.6	91.2	116.5	133.8	136.7
Agricultural	19.2	13.0	14.7	26.8	47.7	33.2	28.8
Manufacturing	49.6	45.6	33.5	60.8	59.8	89.7	92.5
Others	— [a]	2.0	3.4	3.6	9.0	10.9	15.4
Total exports	330.5	272.4	223.1	269.7	334.6	466.0	466.4

SOURCES: 1990–2000 calculated from BCN 2003, tables VI-4, VI-5; 2001–5 calculated from BCN 2007, tables VI-3a, VI-4, VI-5, VI-6, VI-7.
[a]This figure was included in nontraditional agro-exports for 1990.
[b]BCN (2007, table VI-6) includes sugar and beef in its calculation for nontraditional food manufactures, whereas in prior years these two products had been included in calculations for traditional agricultural goods. This author has chosen to include these items in traditional agricultural goods, and the calculations for 2001–5 reflect this.

prises. By the end of 1996, with the privatization process basically completed, 31.4 percent of the land that had composed the state sector had been turned over to the former workers (Cáceres 2003, 120).[23]

A grayer area was what the privatization process would mean for members of cooperatives that were established on land ceded to them by the Sandinistas' agrarian reform. Great insecurity was generated within this sector as a consequence. Ultimately, legislation was agreed on, mandating that most of the land distributed to cooperatives was to remain in their hands. Laws 209 and 210, which were enacted in November 1995, formalized this situation while also calling for the indemnification of many of the former owners (FIDEG 1997). And, Law 278, which was enacted in 1997, provided full recognition of the legality of most of the land distributed by the agrarian reform between 1979 and 1990 (Broegard 2000).[24]

But the reality in the countryside was much more complicated. A major study conducted in 2000 about land tenancy in Nicaragua described numerous situations in which former owners continued to exert pressure to recover their land (Institut de Recherches et d'Applications des Méthodes de Développement [IRAM] 2000b). In some cases, their legal maneuvers had re-

23. In addition, almost 25 percent had been turned over to former members of the army and of the Contra forces; and the rest was returned to its former owners, liquidated, or otherwise dismantled.
24. Yet, IRAM (2000a) implies that these laws left lots of loopholes through which the rights of agrarian reform beneficiaries could be undermined.

1997	1998	1999	2000	2001	2002	2003	2004	2005
331.7	385.1	340.1	408.0	338.7	309.6	320.9	419.9	455.7
244.7	273.7	225.8	266.7	232.0[b]	193.6[b]	208.7[b]	287.5[b]	325.2[b]
87.0	111.4	114.3	141.3	106.7	116.0	112.2	132.4	130.5
245.0	188.1	206.1	234.8	250.7	249.1	290.0	339.9	410.3
79.3	54.7	86.6	90.4	94.7	91.0	104.6	127.7	152.5
153.7	122.3	109.6	133.9	144.4[b]	150.1[b]	178.0[b]	204.7[b]	246.4[b]
12.0	11.1	9.9	10.5	11.6	8.0	7.4	7.5	11.4
576.7	573.2	546.2	642.8	589.4	558.7	610.9	759.8	866.0

sulted in the complete decapitalization of the cooperatives that had been formed by the agrarian reform in the 1980s. In others, the former owners had retaken their farms by force, even though the agrarian reform beneficiaries had legal title to them. Even where the situation had not arrived at either of these extremes, great insecurity reigned. According to IRAM's (2000a, 130) final report, most agrarian reform land was characterized by some kind of tenure insecurity.[25] The threat implied in the privatization process contributed to the massive decollectivization that characterized the sector during the first half of the 1990s (Jonakin 1996).[26]

If one adds the problems generated by economic pressures and bad harvests to insecurity about the ownership of agrarian reform land, one goes a long way toward explaining the sales of this land in the period following the inauguration of the Chamorro government (see Jonakin 1996; IRAM 2000b;

25. Rodríguez Alas (2002) cites a 2001 IICA study that found that 72 percent of landholders in Nicaragua did not possess full title to "their" land. Among this population, agrarian reform beneficiaries experienced the greatest contestation of their property rights.

26. Matus (1994, 7) cites an estimate of 80 percent in the nation as a whole as that part of the cooperative sector that had decollectivized its production by the time he was writing. And, IRAM (2000b, 183) suggests that this process continued apace through the 1990s.

However, not all of those who were sympathetic to the plight of the agrarian reform beneficiaries believed that decollectivization was a negative phenomenon. For example, *Envío* (1994) argued that decollectivization would allow those peasants who were really prepared to meet the challenges facing them to organize their production as they saw fit, working with others to the extent that they considered useful, rather than having to work within a collective structure that was not of their own making.

Broegard 2000; Amador and Ribbink 1993). The initial consequences of these three factors combined could be seen clearly in a study carried out in the Department of Rivas in 1994, which found that approximately 12 percent of the land distributed between 1979 and 1990 had already been sold (Matus L. 1994, 2).[27] The sale of agrarian reform land continued throughout the 1990s; national-level estimates, though admittedly low, were on the order of 32 percent of this acreage having been sold between 1990 and 2000 (IRAM 2000a, 128). In certain parts of the country, especially those with highly desirable land (for its fertility, closeness to infrastructure, etc.), sales were even more common. For example, in the municipality of Malpaisillo—in the Department of León—56 percent of the land that had been distributed to cooperatives in the 1980s had been sold.[28] The same study, quoting another source, estimated that three quarters of the land distributed to cooperatives in Masaya in the 1980s had been sold by 2000.[29] Ruben and Masset (2003, 488), in a study based on two surveys (from 1996 and 1999) analyzing land markets in Nicaragua, conclude that agrarian reform land made up approximately a quarter of the land affected by sales during this period. Agrarian reform beneficiaries who sold all or part of their land listed a variety of reasons for doing so. Jonakin (1996, 1187) found that among those included in his study in northwestern Nicaragua,[30] economic factors ranked highest in terms of reasons for sales. Included here were "'poverty'; the need to acquire 'subsistence'; the absence of 'profits' from agriculture; securing funds to 'pay debts' [many had overdue debts with the state banks, debts whose collection was now being enforced]; or the lack of production credits." Insecurity concerning property rights was the second most cited reason.

This situation was made worse for agrarian reform beneficiaries because their land routinely sold at below-market prices. Broegard (2000, 131) documented this dynamic in the region in which she conducted her case study—Carazo—in the late 1990s. She argues that agrarian reform land was worth approximately half of the value of land received through inheritance. A number of case studies from other regions of the country coincide with Broegard's findings, pointing toward the conclusion that this was a generalized situation (Jonakin 1996; Matus 1994; Merlet et al. 1993). Below-market-

27. In Nicaragua, "departments" are the rough equivalent of province-level jurisdictions.
28. IRAM (2000b)—Section B—Case Study of Malpaisillo and El Jicaral, p. 3.
29. IRAM (2000b)—Section B—Case Study of Masaya, p. 5.
30. Matus L. (1994) and Amador and Ribbink (1993) found the same constellation of factors to be behind land sales by agrarian reform beneficiaries in the regions they studied.

priced sales indicate an inability on the part of the seller to wait for a better offer or a "distress sale." In the surveys from 1996 and 1999, Ruben and Masset (2003, 488) found that distress sales constituted about 40 percent of total land sales. Agrarian reform beneficiaries had a high profile among the cases of such sales. Clearly, circumstances for agrarian reform beneficiaries were far from propitious.

The Overall Effect of SA on Nicaragua's Economy and Its Small Farmers

Nicaragua's small farmers were forced to bear the consequences of SA so that, according to the country's policy makers, the serious economic crisis could be confronted. SA was supposed to bring inflation under control, improve the country's foreign exchange imbalances, reduce or eliminate government budget deficits, and bring about general economic growth. Together these results would enhance Nicaragua's ability to make payments on its foreign debt. Achievement of economic growth would, it was assumed, lead to a reduction in poverty. That is, the newly generated wealth that would be produced by it was expected to "trickle down" through the social structure, improving the lot of even those at the bottom.

The macroeconomic results of the SA were mixed, however. Inflation was reduced from 7,485 percent in 1990 to 10.9 percent in 1995, although it continued to hover around this latter figure through the rest of the 1990s (BCN 2000, table II-3). This was a significant accomplishment. The economic policies were notably less successful, though, in terms of improving Nicaragua's foreign exchange imbalances. The gap between the value of imports and the value of exports remained enormous and the average annual gap only improved slightly: with exports representing an average of 41 percent of imports in the 1980s versus 44 percent in the 1990s (calculated from CEPAL 1997b, table 6.1; BCN 2000, summary page). The less than spectacular growth of exports over the 1990s (discussed earlier and documented in Table 3.3) played a central role here.

With regard to the government's budget deficit, austerity and other measures paid off. The budget deficit continued to be an issue during the 1990s. But it still represented a more positive situation than in the 1980s: the average annual budget deficit in the 1990s represented 9.2 percent of the GDP, as opposed to the 16.5 percent this figure had been in the 1980s.[31]

31. Calculated from CEPAL (1997b, table 6.20; BCN 2000, tables I-1, VII-5, and VII6).

Finally, although the 1990s failed to produce notable growth in Nicaragua's economy, during that period, the negative growth that had characterized the country since 1984 was reversed. Thus, Nicaragua went from having an average negative growth rate of 3.5 percent in the 1980s to an average positive rate of 1 percent in the 1990s (calculated from BCN 2000, table 1.1). Yet given the large annual population increase throughout both of these periods, GDP levels per capita were actually better in the 1980s.

In sum, a few of the SA's principal objectives were met in the 1990s, while others remained somewhat elusive. However, if one places these various indicators in the context of the Contra war ending and the normalization of economic relations with the United States and the international lending community being achieved, the results were much less dramatic than they might initially have appeared. And, they were much less positive than what proponents of SA claim a "return to the market" will bring: the stabilization of key economic indicators, along with economic growth and the benefits typically associated with it.

The SA's results were also much less positive than promised in terms of the effect of these macroeconomic dynamics on the reduction of poverty through a "trickle-down" dynamic. Data from the early period of SA document a notable increase in poverty at the national level: between 1985 and 1993, poverty increased from 42.8 percent to 68.3 percent.[32] Dijkstra (2000, 22) argues, though, that for methodological reasons related to the surveys employed it is best to use data comparing 1993 and 1998. Her data show a slight decrease in poverty, of 2.4 percent,[33] while CEPAL's (2007, tables 4, 5) data show a decrease of 3.7 percent at the national level between these two points in time.[34]

Despite the modest drop in poverty that took place during the 1990s, Dijkstra (2000) also asserts that the results were much less than what might

---

32. See Arana and Rocha (1998, 629). They used the World Bank's $2 (U.S.) per day per capita standard.

Spalding (2007) presents an extensive discussion of different institutions' methodologies for calculating poverty rates in Nicaragua and their consequently distinct figures for this social ill.

33. Dijkstra (2000) is drawing on data that follows the World Bank guidelines that specify that this figure should be calculated taking into account the ability of household income to cover basic consumption needs. According to this methodology, the poverty line was an annual income of $401 (U.S.) per capita in 1998 (see Spalding 2007, n. 7).

34. According to CEPAL (2007, tables 4, 5), Nicaragua had the second highest poverty rate in Latin America. CEPAL relied on national household surveys for their poverty calculations. In these surveys the poverty line was calculated for 1993 at $73.30 (U.S.) per capita per month in urban areas and $49.40 (U.S.) per capita per month for rural areas; for 1998 the corresponding figures were $52.70 (U.S.) for urban areas and $35.50 (U.S.) for rural areas.

have been hoped for if one takes into account that some economic growth took place during this period, and because of the high level of poverty at the outset it should have been easier to reduce it. She attributes this situation to two factors: (1) high population growth, which meant that the absolute number of poor people grew by 9.5 percent, and (2) the increasing inequality of assets. Available data describing changes in income distribution are telling: whereas in 1985 the poorest 20 percent of the population earned just under 6 percent of the income, by 1993, this figure was just under 1 percent (Arana and Rocha 1998, 623). In contrast, the wealthiest 10 percent of the population jumped from earning 27.11 percent of the income in 1985 to 51.49 percent in 1993.[35] Other statistics confirm this trend: the EHII—or Gini equivalent for 1985—was .4161 and for 1993 (and 1998, incidentally), .5590.[36] By early into the new century, Nicaragua had attained the second highest level of inequality in Latin America, the most unequal region in the world (Stok 2005, 69). What occurred after the imposition of SA was not a "trickling down" of wealth but rather a "flooding up" of it.

Looking specifically at the agricultural sector, it was clear that the Chamorro government's SA program, through all of its component parts, had had a significant effect. However, it was not evident that it had been successful in meeting its ostensive objectives in this sector. Increasing foreign exchange earnings through expanded exports had been among the most important of these. However, its movement toward this objective was quite delayed, as well as limited in scope. If one bears in mind the reincorporation of people and acreage into agriculture, following the end of the Contra war, it is not readily apparent that SA was responsible for the expansions that may have occurred in agricultural production during this period. Acevedo Vogl (ca. 2004, 17) notes that the number of people employed in agriculture grew from representing 38.7 percent of the economically active population to 42.6 percent between 1990 and 2000. He argues that this change was entirely responsible for the production increases that occurred during this period. The expanded area under cultivation that resulted from their incorporation

35. According to CEPAL (2007, table 12), the top 10 percent of income earners in 1993 (the earliest data available in their table) garnered 38.4 percent of national income. It rose after that time to 40.6 percent in 2001.

36. The first figure is taken from a University of Texas study measuring income inequality around the world (see http://utip.gov.utexas.edu/data.html), and the second is from the UN University WIDER database (see http://www.wider.unu.edu/wiid/wiid/htm). Reygadas (2006, 121) cites a figure of 0.567 as Nicaragua's Gini coefficient in 1993; and CEPAL (2007, table 14) puts the Gini index at 0.582 for 1993; 0.584 for 1998; and 0.579 for 2001.

into the agricultural labor force was located in the country's agricultural frontier and was characterized by very low levels of productivity.[37] Thus, Acevedo Vogl (ca. 2004, 19) concludes that continued growth of this sort was not replicable because the agricultural frontier was extremely fragile ecologically speaking, the demobilization of soldiers following the war was a one-time phenomenon, and the sector had a restricted capacity to absorb more labor.[38] Hence, a variety of factors came together to generate the growth that took place in Nicaraguan agriculture in the 1990s, with SA policies not being convincingly in the lead among them.

The lack of a dramatic effect of economic liberalization on the generation of export earnings in the Nicaraguan case coincides with findings from Latin America as a whole. Examining the entire region and agriculture in general, Weeks (1995) found no evidence to support the argument that liberalization of the economy, through SA, led to increased production any more than nonliberalization of the economy did. Spoor (2002) takes this a step further in arguing, contrary to the position of SA's proponents, that during the ISI period there had been notable, sustained growth in agricultural production.[39] And, those growth rates have not been regained since the imposition of SA. Even in the area of export agriculture, there was no indication that the countries that had liberalized their economies had surged ahead of the remainder of the region in their production levels (Spoor 2002; see also Weeks 1999).

Yet the earlier discussion also points to the differentiated effect SA had on Nicaragua's agricultural producers. That is, the national-level data suggest that it has been largely negative for small farmers, thus concurring with the literature on Latin America overall. Studies of the region have found that free-market economic policies tend to reproduce or even aggravate inequalities in the countryside (Carter and Barham 1996; Carter, Barham, and Mesbah 1996).[40] Where small farmers are already disadvantaged vis-à-vis agricultural resources before economic liberalization, as is the case throughout most of Latin America, typically it is the larger farmers who are most able to participate in the agricultural growth that occurs because of their privileged access to these resources.

37. Productivity throughout agriculture dropped during the 1990s (Acevedo Vogl ca. 2004).

38. In reaching this conclusion he coincides with a World Bank report from 2000 that he is quoting from.

39. To the extent that earnings from agro-exports dropped during the 1980s—the so-called lost decade—it was due to falling international prices for those goods.

40. De Janvry and Sadoulet (1989), Thrupp (1995), Murray (1998), and Spoor (2002) reach similar conclusions; and Akram-Lodhi et al. (2007) found this dynamic to prevail around the globe.

Conroy, Murray, and Rosset (1996) take an even stronger stance in arguing that the Central American experience in general underlines the numerous problematic effects of the spread of (nontraditional) agro-export production promoted by SA. These included the complete undercutting of the market for the basic grains traditionally grown by small farmers because of the emphasis that had been placed on importing food. Imported food, grown by subsidized producers in the Global North,[41] brought down prices for these goods in local markets. This benefitted urban consumers. But, local farmers also found it difficult, if not impossible, to compete with food aid (see also Gwynne and Kay 1997; Goldin 1993; Edelman 1999).

The remaining alternative for small farmers, according to Conroy et al. (1996), was export crop production. Yet small farmers do not have the same degree of access as large farmers to the resources—credit, technical assistance, and so forth—needed to thrive in the production of the new export crops.[42] Moreover, they do not have the safety net that large farmers have when their crop encounters the selective protectionism of NTAE purchasers or fails for other reasons.[43] Thus, whether small farmers have stayed wedded to basic grain production or have attempted to move into NTAE production, their success has been limited. The logical outcome is a widening of the gulf that separates the production of large "modern" farmers from that of small peasant farmers (see Spoor 2002) and increased rural poverty.[44]

These dynamics also held true for Nicaragua. In Nicaragua, however, the situation was compounded by the more general transformation that the SA program formed a part of—the rapid retreat from socialism toward a complete reinsertion in free-market economics. One expression of the coincidence of these two processes can be seen in the fact that privatization there had a wider effect than in most of the rest of Latin America (here, Chile may be an obvious exception). Because a serious agrarian reform process had

41. As described in Chapter 1.

42. Edelman's (1999) findings on the relationship between Costa Rica's peasants and NTAE coincide with this; Carter and Barham (1996) found this to be the case in other parts of Latin America as well; and Kay (2002 and 2006), likewise, found this to be true in Chile and elsewhere.

43. Rejection of NTAE products by purchasers can occur for a variety of reasons and is not uncommon (Conroy, Murray, and Rosset 1996).

The new export crops also represent a problematic venture from an environmental point of view, as they are heavily reliant on agro-chemicals to control the pests associated with them.

44. See Rello (1999) for a discussion of this issue vis-à-vis Mexico and Central America; Scott (1996) looks at it in Chile, especially during the period of standard SA between 1973 and 1987, while Kay (2002) examines the entire post-1973 context; Nakano (1992) describes this situation in Brazil; and Ganuza et al. (1998) provide an overview of Latin America.

preceded SA in Nicaragua, more agricultural property was in the hands of the state than would otherwise have been the case. Furthermore, the uncertainty surrounding the privatization process led agrarian reform beneficiaries who were not supposed to be directly affected by it (according to the law) to be affected indirectly, given the radically changed political economic environment. The result was decollectivization and, in some cases, sale of agrarian reform land.

The population of those selling land was not only composed of beneficiaries of the Sandinista agrarian reform. Economic pressure resulting from the SA pushed other small- and medium-sized farmers to sell as well. Although using somewhat different-sized categories for groups of producers, Ruben and Masset (2003) found that farmers who fell into the present study's small- and medium-sized group[45] (but excluding those with less than 5 *manzanas*) were the most likely to engage in distress sales.[46] And almost half of all sales during the period they studied (1995–99) were distress sales.

Not surprisingly, most of the studies that have looked at Nicaragua's agrarian land market to date have concluded that a process of reconcentration of land was initiated in 1990 (see, e.g., Ruben and Masset 2003; Amador and Ribbink 1993; IRAM 2000b; and Broegard 2000). In an analysis of the 2001 agricultural census, Baumeister and Fernández (n.d., 15) show that this reconcentration of land did not involve a straightforward shift from the smallest of farmers to the biggest. Rather, the dissolution of the state and collective farm sectors, which had been formed through the Sandinista agrarian reform, translated into an expansion in landholdings in all private-farm-size categories. Yet the main "beneficiaries" of this change were medium- to large-sized farms (see Table 3.4). Together these farms (i.e., those 50 *manzanas* and larger) absorbed 77.9 percent of the land ceded or sold by the state and collective sectors as of 1990. Taking into account this last figure and that the collective and state farms had been composed of small farmers and landless workers, respectively, it becomes apparent that these latter groups lost ground to the capitalist farmer sectors.

In addition to growing land concentration, land sales by poorer sectors of the peasantry also tended to lead to increasing poverty in the countryside. Ruben and Masset (2003, 489) found that "land sales occasioned by distress

---

45. Depending on the region in which they farm, see n. 2 in the Introduction and nn. 1 and 2 in Chapter 4.

46. Interestingly, those with less than five *manzanas* typically did everything possible, including engaging in off-farm wage labor and reducing subsistence, to avoid selling their land.

Table 3.4 Evolution of land tenure in Nicaragua (percentages)

	1978	1988	2001
0–10 *manzanas*	2.1	3.1	4.5
10–50 *manzanas*	15.4	16.7	20.0
50–200 *manzanas*	30.1	28.4	36.6
200–500 *manzanas*	16.2	12.8	18.0
>500 *manzanas*	36.2	13.5	16.5
State land	0.0	11.7	0.4
Collective farms	0.0	13.8	4.0
Total	100.0	100.0	100.0

SOURCE: Baumeister and Fernandez n.d., 15.

reasons caused a fallback to the category of landless for 25–30 percent of the households [that sold land], particularly those in the poor and middle category." And, for the poorer within this group, land sales typically led to a downward shift in their class position and income as they were unable to buy land again if and when the particular crisis that led to its sale passed. Their alternatives then became farming rented land,[47] wage labor—which experienced a dramatic drop in remuneration rates in the agricultural sector in the 1990s[48]—or finding some kind of self-employment.

Thus, distress land sales contributed to the effect that other components of the SA program had on the rural population. Taken together, the SA and its associated processes produced a very high level of rural poverty in Nicaragua. Arana and Rocha (1998, 625) calculate that relative poverty rose from characterizing the lives of 59.3 percent of the rural population in 1985 to 88.7 percent in 1993 and extreme poverty from characterizing the lives of 24.6 percent of the rural population to 52.8 percent between 1985 and 1993.[49]

47. Boucher, Barham, and Carter (2005, 13) show that the landless predominated among households renting land in Nicaragua in the 1990s, with more than 50 percent of renting households being landless.

48. See Acevedo Vogl (ca. 2004, 49).

49. For the purposes of these statistics, the pool of those characterized as living in "extreme poverty" form part of the larger population of people living in "relative poverty." Here, again, this was using the World Bank's $2 (U.S.) per day per capita standard.

Davis and Stampini (2002) posit, in comparing data on poverty from 1998 and 2001, that it had undergone a modest decline during this latter period. Yet they also note that there was an almost frenetic movement at the household level in and out of poverty. In addition, they document the strong relationship between dependence on agricultural production and agricultural wage labor and poverty, essentially arguing that within the currently prevailing economic situation these two sources of livelihood represent a pathway into, not out of, poverty. They call for a variety of institutional supports, including credit and technical assistance, as a means to modify this situation.

Again, following Dijkstra's (2000, 22) admonition that comparing data from 1993 and 1998 is more methodologically sound, CEPAL (2007, table 4) documents a drop in rural poverty during this period, from 82.7 percent to 77.0 percent. But it must be noted that even the latter figure is alarmingly high and that it remained stagnant at that figure through 2001, the latest point for which data are available. It was in rural areas that extreme poverty was especially concentrated (CEPAL 2007; see also Acevedo Vogl, ca. 2004). The lives of Nicaragua's small farmers had clearly been marked by the consequences of exclusionary economic policies, which included increased immiseration.

Conclusion

The wave of neoliberalism that washed over Latin America beginning in the early 1980s brought with it an array of economic policies, especially in the form of SA, that sought to change productive structures and social relations surrounding production in significant ways. The agrarian sector of these countries' economies was not exempted from these changes. Despite abundant evidence that SA programs had fallen far short of the notable production increases they promised to generate, they had an effect on the productive sectors that they excluded from their development agenda—including that of small farmers—and that effect was largely negative.

The Nicaraguan case illustrates the ways in which SA policies negatively affected the small-farmer sector. The austerity measures that reduced their access to key productive resources, the trade liberalization that undercut the market for their products, the currency devaluations that made their inputs more costly, and the privatization that put in question their access to land, individually and together worked to undermine peasant production and, therefore, livelihoods. The situation in Nicaragua was compounded, however, by the larger process of change in which SA was inserted—a rapid retreat from socialism. The shift in the societal balance of forces that this larger process was indicative of led to the various sectors of capital being once again privileged in government policy making over the poorer majority, which included small farmers. This represented a distinct contrast with the redistributive logic that had previously prevailed in policy making there.

Yet the panorama for Nicaragua's small farmers was not entirely bleak. Circumstances permitted some groups of them to take advantage of the new economic environment to their benefit, and others actively sought means by which to soften the blows SA had dealt them. The chapter that follows delineates the many effects of SA on small farmers in four of Nicaragua's rural municipalities. It also suggests a few factors that served to ameliorate the most troubling consequences.

4

The Economic Strategy's Varying Impact
on Nicaragua's Small Farmers

The multiple national-level policies that composed Nicaragua's structural adjustment (SA), which propelled the country's retreat from socialism, were designed to encourage export production. The rationale for emphasizing this sector was that this was the best way to cut down on previously existing inefficiencies in both state and private investments, as well as to ensure the generation of much-needed foreign exchange earnings. Increasing foreign exchange earnings was essential for paying off the country's foreign debt. Meeting that objective was a key goal of both the Chamorro government and its principal advisors in the international lending community. Aside from enabling requests for future loans, payment of the debt would facilitate Nicaragua's full insertion in the world capitalist economy.

The renewed emphasis on export production signaled prioritization of the traditional logic of comparative advantage. Hence, domestically oriented production, and those who engaged in it, were effectively discriminated against by the new policy regime. As described in Chapter 3, this meant that smaller farmers were once again relegated to the back of the bus, as they had been before the Sandinista period. This played out in both the disbursement of state-controlled resources, as well as in taxation policies. Nonetheless, it remained unclear whether the effects of SA policies on domestically oriented producers—especially those engaged in food crop production—were uniform, or whether there was some variation in terms of the extent to which this sector was disadvantaged by them.

To address this question, I conducted interviews with small farmers—who specialized in food production—in four of Nicaragua's rural municipalities in 1997. The municipalities differed in a variety of senses, but perhaps the two most important senses were the degree to which they had been affected by the Contra war in the 1980s and the geographical/ecological makeup of each area. My initial choice of field sites began at the "departmental" (or provincial) level. I decided to study the situation of small farmers in the Department of Matagalpa, which had been directly affected by the Contra war. Because of the heavy presence of small farmers in the municipalities of Esquipulas and San Dionisio, these were selected for the location of my survey (see Map 1).[1] Esquipulas and San Dionisio had been less hard hit by the war than the more northern and eastern municipalities in the department. But, armed bands (the *armados*) continued to be a problem in some parts of these municipalities, and, when possible, even small farmers sought to move their families into the towns for personal security. Yet by and large, farmers there were able to focus their attention on their agricultural production.

The other department chosen as a field site was León, which had not been directly affected by the Contra war. Within this department, the heavy presence of small farmers in Malpaisillo and Santa Rosa del Peñon led me to select them for the location of my survey (see Map 1).[2]

These municipalities also differ, however, on a geographical-ecological level. Esquipulas and San Dionisio are located in the country's central mountain region. Although they form part of the "transitional zone" between the lower, drier areas of Matagalpa (and Nicaragua's western region) and the wetter and higher parts of the department, they are still characterized by their relatively mountainous terrain. In contrast, Malpaisillo is squarely located in the country's Pacific plains region, while Santa Rosa is in the Department of León's northeast foothills. Farther eastward, the foothills area leads into the central mountain region. Given these differences, traditional

---

1. In the Department of Matagalpa, the term *small-scale producer* included farmers with up to thirty-five hectares (or fifty *manzanas*). This definition was taken from Nicaragua's 1952 agricultural census.

2. In the Department of León, the term *small-scale producer* included farmers with up to fourteen hectares of land (or twenty *manzanas*). This definition was taken from Nicaragua's 1952 agricultural census.

—·—·	department boundary
··········	municipality boundary

Municipalities Under Study

1	Malpaisillo
2	San Rosa del Peñon
3	San Dionisio
4	Esquipulas

Map 1   Nicaragua: Four municipalities under study

crop specializations, as well as levels of technology employed in production, varied between the four municipalities.

By choosing such distinct municipalities to include in my study, I hoped to approximate the array of circumstances under which small farmers live and work in Nicaragua. In this way, I sought to discern any differences that might have emerged in the effect of SA on the country's small farmers. As revealed through my findings, the degree to which SA policies affected them varied. Even though most of those interviewed had experienced a notable reduction in their access to productive resources as a consequence of SA, small groupings of farmers had succeeded in either inserting themselves in favorable niches in the newly expanded market or of protecting themselves in other ways from the ravages of it.

These variations were especially clear when comparing the situation of farmers in these two departments. To highlight these differences, the following discussion will be divided into two subsections, each focusing on a separate department. Then, the differences between Matagalpa and León will be analyzed, with special emphasis placed on the causes of them. Hence, the diverse means that Nicaragua's farmers employed in their attempt to survive in the new political economic context, as well as their success or lack of it in this endeavor, will become evident.

Matagalpa

Its higher elevation and cooler climate has made the Department of Matagalpa one of Nicaragua's most important coffee-growing regions for more than a century. In fact, during Nicaragua's first engagement with economic "liberalism," in the late nineteenth century, Matagalpa was a focal point of development and coffee was *the* product of the era. Liberal President Zelaya directed resources toward infrastructural development in that department (and in Jinotega to the northeast) so that inputs could flow in and coffee flow out. And, European immigrants were offered land grants to migrate there if they undertook production of this crop.

Yet a variety of agricultural goods were produced there and continue to be so today, as reflected in the municipalities within this department that were the focus of my study. In Esquipulas, the centerpiece of production for the small farmers I interviewed was the dairy industry. They raised cattle for the dairy products they produced, selling some rustically produced cheese but mostly milk. Basic grains and coffee were grown there to a lesser extent. In contrast, the small farmers I interviewed in San Dionisio concentrated on basic grain production. This municipality is more mountainous than Esquipulas, with many of its rural communities located at a significant distance from its roads. Thus, production was restricted to those goods that were not so perishable as to be affected by its lack of infrastructure. Coffee production and cattle raising were some of the other agricultural activities farmers engaged in.[3]

3. The production patterns of the small farmers included in my survey were entirely representative of those of each of these municipalities (cf. FUNDENIC/INDES 1997, 80–81; AMUNIC 1997a, 1997c).

Despite the attention this department received in the late 1800s, the levels of technology characterizing coffee production, and agriculture as a whole, were low.[4] This was especially true for small farmers. Even in the contemporary period, as seen through the results of my sample, manual production is very common: in Esquipulas, the production of 50 percent of those interviewed was completely manual; and in far-removed San Dionisio, this figure was 93 percent. Irrigation was virtually nonexistent, which kept production costs relatively low. But the produce yields from these farmers' labors were also low.

Even though some differences existed between Esquipulas and San Dionisio in terms of agricultural specialization and technological levels, when compared with the municipalities studied in León, the former two municipalities shared some important commonalities. The small farmers interviewed in the Department of Matagalpa were notably better off than those from the Pacific Coast Department of León. This was evidenced in their access to agricultural resources, as well as by the indicators employed to shed light on changes in their standard of living following the shift of economic policies. For the most part, they appeared to be weathering the storm unleashed by the Chamorro government.

Access to Productive Resources

Among the most crucial issues for farmers worldwide is their access to the resources needed to carry out their production, most especially—but not only—land, credit, and technical assistance. When their access to any one of them is reduced or nonexistent, it limits their productive potential and, therefore, their livelihood. The vast majority of the small farmers interviewed in Matagalpa had secure access to the land they farmed. Most of them (93.3 percent in both Esquipulas and San Dionisio) owned the land they worked (see Table 4.1). They had obtained their land through various means;[5] purchasing was the most common among them (57.1 percent in Esquipulas; and 71.4 percent in San Dionisio).[6] In both Esquipulas and San Dionisio, a noteworthy number of them had gained access to their land

4. See Gariazzo (1984) with regard to coffee on this point.
5. For some of them, their access to land came through several means (e.g., by purchasing one plot and inheriting another).
6. Several of those who had purchased their land had done so from former beneficiaries of the Sandinista agrarian reform.

Table 4.1   A comparison of four municipalities: Changing access to agricultural resources and its effects on small farmers (percentages)

	Matagalpa		León	
	Esquipulas (N = 15)	San Dionisio (N = 15)	Malpaisillo (N = 15)	Santa Rosa (N = 15)
Land				
Owned	93.3	93.3	46.7	73.3
Purchased	57.1	71.4	57.1	54.6
Inherited	21.4	21.4	42.9	54.6
Agrarian reform	35.7	35.7	28.6	18.2
Credit				
Before 1990	100.0[a]	91.7[a]	66.7[a]	27.3[a]
In 1997	28.6[b]	53.3[b]	6.7[b]	13.3[b]
BANADES	7.1	—	—	—
NGO	21.4	53.3	6.7	13.3
Technical assistance				
Before 1990	83.4[a]	84.6[a]	44.4[a]	33.3[a]
In 1997	21.4[b]	46.7[b]	26.7[b]	6.7[b]
Changing production patterns				
Increased production w/ credit	50.0	46.7	6.7	—
Increased production w/o credit	28.6	13.3	20.0	20.0
Reduced production/lack of credit/debt	21.4	6.7	13.3	—
Improvements on the farm				
Were possible	80.0	66.7	26.7	66.7
Purchases of animals/ equipment	66.7	73.3	26.7	40.0
Forced sales of animals/ equipment	6.7	6.7	20.0	20.0
Member of an organization				
Yes	57.1	80.0	13.3	73.3

SOURCE: Author's survey data.
[a] Here the denominator is the number of people who were farmers before 1990.
[b] Here the denominator is the number of people who were farmers at the time of the interview.

through the Sandinista agrarian reform.[7] The remaining means by which they attained their land was inheritance.

These farmers were relatively fortunate vis-à-vis their access to land, but this was less true of their access to formal credit after 1990.[8] Before 1990, most of them who had been farmers had attained credit through the Na-

7. One of these agrarian reform beneficiaries from Esquipulas sold all of his land in the 1990s (ostensibly because of the threat to his production and personal safety implied by the continuing presence of *armados* in the vicinity of his farm) and was landless at the time of our interview.

8. That is, credit made available from a lending institution or an NGO with a lending program.

tional Development Bank (BANADES). In contrast, in Esquipulas only 28.6 percent received credit in 1997, while in San Dionisio, 53.3 percent did (see Table 4.1).[9] Of equal importance, only one of these thirty informants received credit from BANADES that year. Nongovernmental organizations (NGOs) were the source of credit for the remainder of those who had been able to obtain it.

During the Sandinista period, most of these small farmers had also received technical assistance from the government. Yet this resource had become even harder for them to attain than credit in the post-1990 period. In contrast to the 83–84 percent who received technical assistance in the past, only 21.4 percent of those interviewed in Esquipulas, and 46.7 percent of these interviewed in San Dionisio received it in 1997 (see Table 4.1).[10] And, *none* of those who were fortunate enough to receive it in the mid-1990s could look to the government for it. Instead, in almost all of the cases, it was NGOs that now provided technical assistance.

The Effect of Reduced Access to Productive Resources

Given the major reduction in access to agricultural resources experienced by these farmers, it is essential to assess its effect on their production. Perhaps the most remarkable change was that a larger number of them than might have been expected had been able to increase the size of their cattle herd, or expand their cultivated area. They did so with the support of a loan from an NGO or state institution. Fifty percent of those interviewed from Esquipulas and 46.7 percent from San Dionisio had accomplished this (see Table 4.1). Of equal significance, several informants from each municipality had expanded their herd size or area under cultivation without any apparent source of outside assistance. Fifty percent of the small farmers interviewed in Esquipulas had even been able to plant improved grasses for their cattle to feed on. Moreover, fewer of them than might be expected had been forced to cut back their herd size or area under cultivation because of lack of credit.[11]

Another key indicator of the economic strength of their farms was the

9. This notable difference between Esquipulas and San Dionisio will be addressed in the text.

10. This difference between Esquipulas and San Dionisio will also be addressed in the text.

11. For only two of those interviewed in Matagalpa (who both lived in San Dionisio) was their production of a mere subsistence nature (i.e., with them having made no sales of their produce from the previous agricultural cycle).

extent to which they could provide employment for these farmers and their families.[12] While in neither of these two northern municipalities were all of those interviewed employed full time on their farms, a notable number of them were. In Esquipulas, those who worked full time on their farms represented 64.3 percent of the sample. The most important secondary source of income was various forms of self-employment.[13] In contrast, San Dionisio was characterized by fewer of the informants working full time on their farms (46.7 percent). However, another 26.7 percent of these were engaged in working with peasant organizations. This was typically not an activity engaged in out of economic necessity but rather out of a commitment to bettering the lives of those in the community. The remainder engaged in agricultural labor, the urban labor force, or self-employment.

Furthermore, their farms also served as a source of employment for others. In Esquipulas, 50 percent of these peasants had at least one son employed on their farm (most of their wives were housewives). In San Dionisio, 53.3 percent of their spouses worked at least part of the time on the farm, while the same percentage of farmers had sons working on the farm. In addition, work on most of the farms required the assistance of hired temporary or permanent laborers.[14] Thus, their farms were strong enough economically to sustain many of these peasants with farm-generated income alone, as well as to provide some family members and hired laborers with employment.

Several other indicators also point to the economic situation of these farmers, including the extent to which they had been able to make improvements on their farms in the post-1990 period. Eighty percent of those interviewed in Esquipulas and 66.7 percent of those in San Dionisio had been able to do so (see Table 4.1). This was quite noteworthy given the difficult economic circumstances characterizing many sectors of the population at the time. Likewise, the majority of them had been able to purchase either farm animals or equipment after 1990.[15]

Yet, some of the small farmers I interviewed in Matagalpa were experiencing economic hardship. They relied on various strategies to "get by,"

---

12. The official unemployment rate in Esquipulas in 1995 was 11.2 percent and in San Dionisio, 11.6 percent (calculated from INEC 1997, 329).

13. Only one of these farmers also sold his labor power in agriculture.

14. This is in addition to members of their nuclear and extended family who worked on the farm.

15. In contrast, only 6.7 percent of the informants in each municipality had been forced to sell either farm animals or equipment, without having the ability to replace them.

including working their farms more to generate additional income, looking for work off the farm, and taking out informal loans. But, even among those who felt that their economic situation was less than ideal, most were able to look to the farm as a solution to their problems.

León

The two municipalities in León where I interviewed small farmers were quite different from each other. The first, Malpaisillo, historically had fertile land that had been totally consumed by cotton production since the early 1950s. As a municipality, it had produced most of the region's cotton (AMUNIC 1997b, 15). Cotton production tended to be the domain of large producers (see Enríquez 1991). Although the crop was "rain fed" rather than irrigated, its cultivation was largely mechanized. But decades of monocultural production had depleted the fertility of the soils and contaminated the water tables. When the international price for cotton plummeted in the early 1990s, it was partially replaced by sesame, yucca, sorghum, corn, and beans (only the first two of which were produced for export). Despite this shift in emphasis, technological levels remained higher there than in other areas where the focus was on domestic crops. Among those interviewed in Malpaisillo, animal traction was employed by 86.7 percent. And, a full 40 percent also relied on tractors (rented or otherwise) in at least some part of the production process.

In contrast, corn and bean production prevailed in Santa Rosa del Peñon.[16] True to the national pattern, technological levels employed in producing them were quite low. Among the small farmers I interviewed, manual production was predominant (86.7 percent). But, animal traction was also used by 33.3 percent of these farmers in at least some of their production.

Yet Santa Rosa had also contributed to León's cotton production in the past, as a supplier of temporary labor for the harvests. This region had also had a mining heyday, when its gold mine—El Limón—had offered employment to many of its wage laborers. In the late 1990s, however, given the agricultural stagnation and the overall economic crisis in the area, many laborers worked individually in the mining of gold, gypsum, and lime

16. The small farmers included in my survey were entirely representative in their production patterns for each of these municipalities (AMUNIC 1997b, 1997d).

(AMUNIC 1997d).[17] This municipality was also noted for its deforested hill-sides, sparse rainfall, and extreme lack of infrastructure in comparison with the rest of León.[18]

Access to Productive Resources

There were some additional differences between these two municipalities, which included the degree of access to land of the small farmers in each. In Malpaisillo, only 46.7 percent of those interviewed had secure access to land through their ownership of it. In Santa Rosa, 73.3 percent of them owned their land (see Table 4.1). Those farmers who owned the land they worked came by it through several means. In Malpaisillo, more of these landowners had purchased their land than inherited it, while in Santa Rosa an equal number of farmers had become landowners by inheriting their land as those who purchased it.

The Sandinista agrarian reform had not had as much effect on the farm-ers interviewed for this study in León as it had in Matagalpa. Nonetheless, it is clear from the research of others that those who had been agrarian reform beneficiaries in León—and of these two municipalities, especially those in Malpaisillo—found their access to that land seriously threatened by former owners in the 1990s (Institut de Recherches et d'Applications des Méthodes de Développement [IRAM] 2000b). The threats took various forms, ranging from violence against cooperative members to mounting law-suits to reclaim the land that had been theirs previously, which required lengthy and costly legal expenditures for the agrarian reform beneficiaries. Moreover, the legal battles took place largely within a system that was un-sympathetic to those who had received land through the agrarian reform. Undoubtedly, it was the high productive potential of the land in the depart-ment—particularly in the plains area, including Malpaisillo—that led to the conflict being so intense there (see IRAM 2000b). As noted in Chapter 3, the insecurity caused by the actions of former owners contributed signifi-cantly to the decision by cooperative members to sell their land, a dynamic that had affected more than half of the land in Malpaisillo that had been

17. The agricultural stagnation was caused in part by lack of rainfall. The economic crisis of this period had even pushed children into work in the mines (see *La Prensa* 2001).

18. According to CONAGRO/Banco Mundial (1995, 20–21), only 3 percent of the commu-nities in Santa Rosa were accessible by vehicle year round, in comparison with the overall average of 43 percent for León.

owned by cooperatives formed through the Sandinista agrarian reform (IRAM 2000b, 9).

However, economic pressures during this period had taken away some of the land that had previously been accessible to the small farmers I interviewed. Two of the five farmers in these municipalities who had sold their land in the post-1990 period had been forced to do so specifically because of economic necessity. A few of those who engaged in land sales were left with no land of their own. In both municipalities, all of the landless farmers interviewed engaged in some type of rental arrangement to obtain access to it.[19]

These farmers' access to agricultural credit was even more limited than their secure access to land (see Table 4.1). Whereas 66.7 percent of those interviewed in Malpaisillo who had been farmers prior to 1990 had regular access to credit from BANADES, in 1997 only 6.7 percent did. In Santa Rosa notably fewer of those interviewed who had been farmers had attained credit prior to 1990 (only 27.3 percent), but their access to this resource in 1997 was also very restricted (13.3 percent). In both municipalities, to the extent that these farmers were able to attain credit in 1997, it was NGOs that provided it. Not one of them received it from BANADES.

In some cases, the lack of credit translated into an inability to purchase any of the inputs normally employed in agricultural production. Hence, no fertilizers or pesticides were employed in production. Seeds were borrowed or kept from the harvest of the previous year to provide the starting point for the next crop. In Malpaisillo, this was so for 26.7 percent of the sample, and in Santa Rosa, 13.3 percent of the sample. Were it not for the area's depleted soils, and the lack of an integrated plan for the renovation of their fertility, this approach might eventually have facilitated the transition to high-value organic agriculture. Yet given the situation of the majority of Malpaisillo's small farmers, the absence of inputs instead represented a step backward toward mere subsistence production.

Technical assistance was similarly circumscribed for these small farmers in León. While their receipt of this resource even before 1990 had been limited (44.4 percent in Malpaisillo and 33.3 percent in Santa Rosa), it was even more restricted in the 1990s (see Table 4.1). In Malpaisillo, 26.7 percent had received it in the previous agricultural cycle (1996) and all of them had

---

19. However, IRAM (2000b) found that many former agrarian reform beneficiaries who lost all or some of their land in the 1990s had been forced to emigrate in search of employment, to Managua or even to as faraway as Costa Rica.

to pay for this service. In Santa Rosa, only 6.7 percent had received it. In both municipalities, NGOs had provided this service for all of those who were fortunate enough to receive it.

In the post-1990 period, technical assistance was even more important in León than elsewhere because most of the department's farmers had been forced to replace their cotton crop with other crops. Thus, they had to select the most appropriate replacements for it and become acquainted with these new crops' production processes. The absence of this assistance undoubtedly made the changes in production even more difficult.

The Effect of Reduced Access to Productive Resources

The reduced access to agricultural resources affected the production patterns of León's small farmers. In Malpaisillo, where the cotton regime had collapsed, more than half of these small farmers had made changes in their cropping patterns. In both municipalities, though, only 30 percent had adopted the cultivation of new crops (i.e., as opposed to simply cutting back on the number of crops they produced, while continuing to grow some traditional crops). Of the minority who had been able to expand their area under production, most had done so without any kind of formal financial assistance. At the same time, a small number of farmers in Malpaisillo had specifically had to cut back their production because of lack of official credit, or the loss of their land stemming from their debt to the bank. But, only in Santa Rosa, and only among 20 percent of the farmers interviewed there, did production levels drop so low that none of what was harvested was marketed.

How did changes in production and reduced access to agricultural resources affect these farmers' living standards? In neither municipality were many of the farmers able to live on their farm-generated income alone.[20] In Malpaisillo, 46.7 percent were employed full time on their farms, while in Santa Rosa, only 40 percent were. Most of those who had to supplement their farm-generated earnings, did so by participating in the agricultural labor force (62.5 percent in Malpaisillo and 88.9 percent in Santa Rosa). Other sources of income were various kinds of self-employment and urban wage labor.

However, many of their farms also absorbed the labor of other family members or hired workers. Seventy-three percent of those interviewed in

20. The official unemployment rate in Malpaisillo in 1995 was 20.6 percent and in Santa Rosa, 11.6 percent (calculated from INEC 1997, 315).

Malpaisillo had at least one son assisting them on the farm, and 46.7 percent did in Santa Rosa. They also offered temporary employment to hired laborers. What these figures suggest is that, by and large, their farms were unable to generate enough income to employ either the informants or their relations full time but that there were peak periods of activity that required them to turn to relatives (their sons and extended family) and hired labor for assistance.

Another indicator of how these farms were faring was the extent to which their owners had been able to make improvements on them. Here there is a striking difference between the two municipalities: only 26.7 percent had been able to make improvements in Malpaisillo, in contrast with 66.7 percent in Santa Rosa (see Table 4.1). Although a fair number of the informants were able to purchase farm animals or equipment, 20 percent in each municipality had been forced to sell some farm animals or equipment without being able to replace them.

Among those who were having difficulties getting by, a number of strategies were employed in their attempt to cope. In Malpaisillo finding work off the farm was the means most employed by those in need. In Santa Rosa, where the need was less extreme, farmers took out informal loans or simply consumed less.

Matagalpa Versus León

The contrasts between the two municipalities within each department were relatively small compared with those between the departments. Going from Esquipulas to San Dionisio in Matagalpa, to Santa Rosa, and, finally, to Malpaisillo in León, these farmers were in decreasing numbers owners of the land they farmed. Hence, farmers in Matagalpa had more secure access to land than those from León (especially Malpaisillo). And, the informants in Matagalpa also had notably more access to agricultural credit than those from León. Finally, small farmers in Matagalpa were similarly at an advantage in relation to those from León in their access to free technical assistance.

Differential access to resources was reflected in these farmers' changing production patterns. In both Esquipulas and San Dionisio, a sizable number of these farmers had been able to increase either their production or their cattle herd size with the credit they received. This was significantly less true of the informants in León. If farmers who had been able to make purchases of farm equipment or animals were brought into the equation, the municipalities in Matagalpa were once again at an advantage over those in León. The varying number of informants who had been forced to sell some equipment or farm animals without having the ability to replace them also highlights that those in León were in a more difficult economic position than those from Matagalpa.

These patterns reveal that at the time of my survey in 1997 small farmers in the Department of Matagalpa (as evidenced by Esquipulas and San Dionisio) were experiencing less economic hardship, in the context of SA, than those from León (as evidenced by Malpaisillo and Santa Rosa). However, by the end of the 1990s, the Department of Matagalpa was the scene of much misery, as many small farmers and agricultural workers who depended on earnings from coffee production experienced a dramatic downturn in their economic situation. In 1999, the international price of coffee was approximately 26 percent less than it had been in 1998, before falling another 10 percent in value in 2000 and an additional 25 percent in 2001 (BCN 2003, table VI-8). Many coffee farmers stopped all maintenance of their plantations, and some even let the ripe crop fall to the ground, calculating that they would not earn enough from the harvest to cover their production costs. Road blocks were set up in many locations in the coffee-growing regions, as whole families "came down from the mountains" to demand economic assistance from the government, and the World Food Organization acknowledged that people in this region were starving.[21]

Although none of the farmers I interviewed were primarily coffee producers, and even though my survey was carried out before this coffee "bust" period, economic issues were at the forefront of all of their minds when they were asked about the most serious problems currently confronting Nicaragua's small farmers. This was the case in both León and Matagalpa. They all pointed to lack of agricultural credit as the greatest problem faced by small farmers. The high cost of production was raised by a number of these farmers as well. Some of the informants also mentioned the low prices paid for produce. Clearly, the economics of farmers' livelihoods were of concern to them.

Despite the many problems my interviewees considered small farmers to be confronting, it was evident that they, themselves, were not all affected

21. See, for example, *El Nuevo Diario* (2001) and González (2001).

to the same extent. Yet most of the government's economic policies were implemented nationwide. Thus, it remains to be explained why these policies had such a differing effect between the two regions. It is to this puzzle that I now turn.

The Sources of These Regional Differences

These findings illustrate the extent to which the small farmers I interviewed had been affected by the structural adjustment of Nicaragua's economy that had formed the centerpiece of the retreat from socialism the Chamorro government set in motion. These farmers had found their access to a number of key agricultural resources severely cut back from 1980s levels. For many of those interviewed, the growing restrictiveness of formerly available resources meant that they produced less, reduced their consumption, or found other means of supplementing their farm-generated income. In the worst cases, these farmers were forced by debt to sell part or all of their land and joined the pool of the landless.

Nonetheless, there was a sizable group of the farmers I interviewed whose economic situation had been less negatively affected than was the norm by the government's new policy orientation. This was expressed in terms of their relatively greater access to credit, technical assistance, and so forth. But it was also indicated by their ability to expand their production and/or to bring about improvements on their farms or homes. The factors distinguishing these more privileged farmers from the rest of their counterparts were (1) what they produced and/or (2) their participation in an organization that provided them with at least some of the resources they had previously received from the government.

SPECIALIZING IN THE "RIGHT" PRODUCT

In Esquipulas and, to a lesser extent, in San Dionisio, the small farmers I interviewed were engaged in raising dairy cattle. Although cattle raising is not a new economic activity in Nicaragua, having undergone a period of great expansion between the late 1950s and 1980, a couple of changes had taken place as of 1990 that had a significant effect on it. The first was that the Contra war came to an end, which allowed cattle raising to expand again. The war had largely been confined to the central mountain region, the same region where most of the country's cattle herd was raised. The return of peace to this region, even though not instantaneously or uniformly achieved, meant that areas formerly unsafe for cattle raising were available once again, and the herd could undergo expansion. This was a welcome respite, given that during the war against Somoza—especially the last few years of it—as well as the Contra war, the nation's herd had experienced serious losses.

In addition, some new avenues for marketing dairy products had been opened up. The experiences of the dairy farmers I interviewed illustrated these well. Some of these farmers sold their milk and by-products to private companies that sent trucks into the region each day. Others delivered their milk directly to a local plant that was established by a group of farmers. Before setting up their own plant (the Agricultural Credit and Service Co-operative, R. L., Esquipulas) this latter group of farmers sold their milk to a private company (PROLACSA) based in the department capital, Mata-galpa.[22] Because of milk's perishable nature, they were completely dependent on PROLACSA to pick up and transport it in a timely fashion. This put them in a vulnerable position regarding decisions about the price they received for their milk. Thus, these farmers organized themselves into a coop-erative and applied for financial assistance from a government development program that had foreign financing. With this funding, and their own mem-bership fees, they were able to set up their own milk plant in 1995. The plant's establishment allowed them to manage the commercialization of their product. The plant functioned as a refrigerated storage center. In set-ting it up, they cut out at least one level of intermediaries, thereby allowing more of the profits to remain with them. Having their own plant also meant that they had more control in negotiations over prices, as they no longer relied on others to ensure that their milk reached refrigeration facilities be-fore going bad. Had this plant also pasteurized their milk, their position would have been even stronger, but the plant still represented a crucial step forward for the producers. These types of plants, as well as those engaged in milk processing on a small scale, were becoming more common throughout the cattle-raising region.

At the same time, the milk plant and its members benefited from the new export market that had opened up for Nicaraguan milk and cheese. Most of Nicaragua's milk products continued to be sold to the domestic market.[23]

---

22. PROLACSA, a subsidiary of Nestlé, is the second largest milk-processing company in the country (IICA 2003, 112). All of the milk it processes is converted into powdered milk destined for the local, as well as the export, market.

23. One estimate put the export/domestic market breakdown at 25 percent/75 percent in 2001 (Interview, UNAG, May 14, 2001).

But, the existence of an export market meant that the demand for such products underwent an expansion and that their producers had additional options in terms of where they could market their milk and cheese. This new market consisted of buyers from El Salvador. Given its relatively large population and the small size of its national territory, which limited the space for pasturage, El Salvador's consumption of dairy products was heavily dependent on imports from other countries.[24] Nicaragua's milk supply was tapped to meet El Salvador's need for dairy products and by 1998 El Salvador had become Nicaragua's biggest market for milk exports (*La Prensa* 1999). Salvadoran buyers purchased milk and cheese produced throughout the corridor beginning in Esquipulas and moving southeast through Boaco and Chontales (as least as far as Nueva Guinea).

This marketing relationship had existed before 1979. But it fell apart early on in the period of Sandinista governing as the new government stepped in to control the international marketing of Nicaraguan produce. Like its efforts to control domestic commercialization of produce, the stated objectives of this policy had been to ensure a stable price for domestic producers (i.e., to protect them from vacillations in international markets), guarantee access—at reasonable prices—for local consumers, and, if there was a surplus generated by a surge in international prices, to channel it toward domestic social or development projects rather than individual luxury consumption by producers.[25] State control over international marketing was part of the Sandinista model of development. However, because of the major differences in political economic orientation between the Sandinistas and the rest of Central America's governments in the 1980s, serious tensions emerged between them. Those tensions contributed to the virtual breakdown in some areas of trade between them, including that of the Salvadoran marketing network for Nicaragua's dairy products.

With the change in government in 1990 came the opening up of Nicaragua's borders for increased foreign trade. The lifting of government controls on trade had a mixed effect on producers, hurting some and bringing gains to others. Most of Nicaragua's small farmers were first and foremost basic grains producers, and they were negatively affected by the influx of food aid

24. El Salvador also re-exported part of its Nicaraguan milk imports to the United States. That is, it used part of its U.S. milk export quotas for milk products it imported from Nicaragua (Interview, UNAG, May 14, 2001).

25. On the logic behind state control over international and domestic marketing of produce, and some of the dilemmas it generated, see Utting (1987), Spoor et al. (1987), and Irvin (1983).

and imports. A variety of other goods, and their producers, were likewise hit by the influx of cheaper products that competed directly with them.

Yet producers of milk and its by-products were able to take advantage of the new opportunities for marketing their produce and to reap notable benefits from them. And take advantage of them they did. According to *La Tribuna* (1996), milk exports almost tripled between 1995 and 1996 alone. These figures also reflected the overall growth in the importance of this sector for the economy. The cattle industry, which produces livestock, meat, milk, and cheese, generated approximately 7 percent of Nicaragua's gross domestic product during the 1990s (IICA 2003, 9).[26] This was more than any other agricultural product. And small- and medium-sized producers were responsible for the bulk of dairy production (Flores Cruz and Artola 2004).

Benefits for farmers were especially to be had from dual-purpose livestock raising (i.e., for milk and meat). Clemens (1994, 14–15) found that this was one of Nicaraguan agriculture's most competitive sectors in the early 1990s.[27] In comparing two recent rural household surveys, Davis and Stampini (2002, 16) argue that cattle raising represented a key "pathway out of poverty" for those in the agricultural sector of the population. Moreover, accounting for production costs and yields, the area of the country that Matagalpa is located within (according to delimitations employed in the 1980s and early 1990s, Region VI) was second only to that containing Boaco, Chontales, and Nueva Guinea in terms of the profitability of milk production. Thus, Esquipulas's and San Dionisio's milk producers found themselves in an unusually advantaged position in the 1990s.[28]

At the other extreme of the continuum represented by these four municipalities was Malpaisillo. The area's principal crop until the early 1990s was

26. And, Flores Cruz and Artola (2004, 151) attribute 33.5 percent of the agricultural GDP to the dairy industry.

27. A high-ranking official from the livestock sector of the National Union of Farmers and Ranchers (UNAG) argued that milk production was not profitable in Nicaragua, at least at the time of our interview, which was conducted seven years after Clemens's study was completed (Interview, May 14, 2001). His critique of government policy toward this sector was echoed in a number of studies about milk commercialization from the first half of the 1990s (e.g., Cajina Loáisiga 1993; Análisis Total 1995).

28. In the late 1990s, yet another major actor entered Nicaragua's milk commercialization network, the Italian dairy corporation Parmalat. In contrast to a few of the other large dairy product companies in Nicaragua, it depended fundamentally on locally produced liquid milk (as opposed to imported powdered milk) for its processing. And, by 2003 Parmalat had become the largest processor of Nicaraguan milk (IICA 2003, 112), which it sold both in the local and international markets. Parmalat's entry into the country's dairy market opened up even further the avenues through which dairy farmers could sell their milk.

cotton. At that time, its price in the international market dropped significantly.[29] But, its production costs remained high because of its heavy dependence on imported inputs. The result was that it became impossible to earn a profit from cotton production using prevailing technologies (cf. Clemens 1994, 16). As a consequence, its cultivation virtually disappeared from the regions that it had dominated from 1950 to 1990. The entire Pacific Coast region of Guatemala, El Salvador, and Nicaragua, where it had been so central to the economy, experienced the effects of its demise.

For cotton farmers, including some of those interviewed as part of the survey, this meant that they were abruptly forced to find alternative crops. But the disappearance of cotton from Malpaisillo, and other municipalities in this region, also meant that the labor market for workers who assisted in its cultivation, and especially its harvesting, likewise dried up.[30] Hence, Santa Rosa's small farmers were included among those hurt by the demise of cotton. *Tiempos del Mundo* (2000) described the situation in León and Chinandega as being "on the edge of an economic emergency," with unemployment running as high as 80 percent. The extraordinarily high level of unemployment, in large part caused by the region's agricultural crisis (especially that of cotton production) had numerous spillover effects in other economic sectors, bringing down many commercial and industrial enterprises with it. From one year to the next, an entire regional economy was thrown into disarray when the bottom dropped out of the cotton market.

Ironically, the international market had showered many dairy farmers in Matagalpa with benefits, while it wreaked devastation in León. Cotton production had always depended on the international market, with its booms and busts. At least for the moment, the opening up of Nicaragua's market in dairy products had brought about a boom for its producers.

"COUNTERMOVEMENT" IN THE FORM OF FARMER ORGANIZATIONS
Good fortune with a principal product in the international market, however, does not explain all of the cases in which SA had not weakened the eco-

29. In fact, the international price for cotton had been even lower at several points in the 1980s (see Clemens 1994, Anexo 7).

30. According to *Envío* (1993, 17), at the end of the 1980s and very beginning of the 1990s, Nicaragua's cotton harvest had generated approximately 70,000 jobs.

IOM/INEC/UNFPA (1997) discusses the relationship between the demise of cotton, the loss of harvest jobs, and the resulting emigration of rural dwellers out of the municipalities where it had been produced, as well as out of the surrounding municipalities whose populations had also relied upon its wages.

nomic position of the small farmers I interviewed. The additional explanatory factor was the extent to which these farmers participated in local organizations that facilitated their access to at least some of the services that they had previously received from the state or that aided them in some other way within the new economic context.[31] Even in the case of Esquipulas, where the product they produced was a key variable in the equation, organization among these farmers was also important. Clearly, the organizations that many of them (57.1 percent) formed a part of, including the Agricultural Credit and Service Cooperative, R. L., Esquipulas, had put them in a position to take advantage of a positive situation in the market for their product.

But the situation of small farmers in San Dionisio provides the best example of the crucial role that organization among them played. There, most of my informants followed the pattern set throughout the municipality by specializing in the production of basic grains. Nonetheless, basic grains were not characterized by particularly good prices in the 1990s. Because basic grains were largely destined for the domestic market, their producers were at the low end of the list of sectors to receive the increasingly scarce government resources designated for the agricultural sector. Thus, the products these farmers focused their agricultural labors on were not what served to ameliorate the negative effects of SA.

Instead, organization was these farmers' salvation. In San Dionisio, which is noted for the very high level of mobilization of its rural population, 80 percent of those interviewed had worked for at least a few years during the 1990–97 period with some kind of group that brought farmers together (see Table 4.1). The one of most consequence there was known as the Asociación Indígena, which was founded in 1990. Its name underlines the considerable presence of an indigenous population in the region. The Asociación was established to promote the economic, social, and cultural development of this indigenous community. Another organization that a fair number of my informants were members of was the Banco de Granos de San Dionisio/ PRODESSA. PRODESSA was actually formed in 1986, even earlier than the Asociación Indígena, and carried out projects in various parts of the Department of Matagalpa (Espinoza 1994). The Banco de Granos, which worked in conjunction with PRODESSA, was established in 1993. These

31. The study of Davis and Stampini (2002) also stressed the importance of participation in producers' organizations as a means of alleviating the difficulties—stemming from the lack of institutional supports for agricultural production—characterizing this sector in the 1990s. See also Davis, Carletto, and Sil (1997) in this regard.

two organizations, along with a few more associated groups, specifically focused on meeting the needs of the area's small farmers.

These organizations supplied the farmers who worked with them some of the resources they were cut off from with the initiation of the Chamorro government's SA program. They provided them with some credit—although typically not as much as they had previously received from BANADES—technical assistance, training in agriculturally related skills (e.g., soil conservation), and assistance with the storage and marketing of their grains to protect them from the extreme vacillations of the grain market.[32]

In the 1990s, a multitude of organizations sprang up in Nicaragua, most of which sought to bring together and improve the position of the less-advantaged sectors of the population. These groups, whether newly formed or with recently expanded activities, were often the recipients of funding from foreign NGOs and governments seeking to foster the growth of the private sector and/or civil society. Inevitably, farmers who were organized were much more likely than those who were not to receive the benefits that might come from these international relationships for the simple reason that, from the point of view of the funders, their organization facilitated the implementation of projects. Thus, San Dionisio's previous history of having a highly mobilized rural population definitely worked in its favor, making the heavy presence of NGOs with projects there of little coincidence.

Despite not having a pre-1990 history of being significantly more organized than other municipalities, Santa Rosa's rural population had been thought to have the potential to become organized.[33] It was assumed this was due to the extreme poverty that had characterized Santa Rosa for decades (this was the poorest of León's municipalities) as well as the strong training in organizational development that the National Union of Farmers and Ranchers, the Agricultural Workers' Union, and the Sandinista National Liberation Front had engaged in there during the 1980s.[34] Whatever its origins, this potential was evidently borne out because a remarkable number (73.3 percent) of those included in the survey from Santa Rosa had partici-

32. See Jonakin and Enríquez (1999) for further information concerning the role played by the nontraditional financial sector in Nicaragua in the 1990s and Bebbington and Farrington (1993) on the ways in which NGOs fill gaps in government programs in the wake of SA throughout the Global South more generally.

33. Interview, Rural Development Consultant, who worked throughout most of the 1980s for the Ministry of Agricultural Development and Agrarian Reform, March 17, 1998.

34. Interview, UNAG activist, July 16, 1999. This informant had worked in the municipality of Santa Rosa at various times since the early 1980s.

pated in some kind of group working in the municipality's rural areas during at least a few years in the early 1990s.

It is essential to note a key difference between the organizations working in San Dionisio and those in Santa Rosa. In contrast to farmers in San Dionisio, those in Santa Rosa generally did not receive credit and technical assistance from the groups there. The principal assistance organizations working in Santa Rosa offered was training in soil conservation techniques. Familiarity with these techniques was useful, given that this region was considered an "ecological disaster area" because of the massive deforestation that had occurred in recent decades, which resulted in serious soil erosion problems and dramatically reduced rainfalls. Clearly, all of this had an effect on farming in the area. So attaining skills of this kind could assist the municipality's small farmers to work toward reversing (or at least ameliorating) some of the damage wrought by deforestation. Nonetheless, acquiring knowledge about soil conservation techniques was not the same as receiving agricultural credit or even technical assistance. It could not be translated so immediately and directly into economic benefits for program participants.

At the same time, the serious lack of infrastructure in Santa Rosa would have also worked against farmers in this department. It would have complicated their attainment of inputs and, even more important, the marketing of their produce. Even if they were producing the same crops as farmers elsewhere, such as those in San Dionisio, their isolation put them in a much less advantageous position regarding the market.

Hence, several of the indicators used to suggest standard-of-living changes that had characterized these farmers since 1990 point to a difference between these two highly organized municipalities. For example, fewer of Santa Rosa's informants had been able to make purchases of farm equipment or animals than their counterparts in San Dionisio (40 percent vs. 73.3 percent, respectively). And, the forced sale of agricultural equipment and animals was somewhat more common in Santa Rosa than in San Dionisio (20 percent vs. 6.7 percent, respectively). Yet the organizational network that had been brought into existence in Santa Rosa contained within it the potential for other NGOs to enter the area and bring different kinds of resources, including credit and infrastructure, which might be more immediately beneficial in economic terms.

At the other end of the organizational spectrum was Malpaisillo. Only a few (13.3 percent) of the small farmers that I interviewed there worked with some kind of group. Regardless of the reasons leading Malpaisillo's small farmers to participate less in rural organizations than small farmers in the other three municipalities, the consequences for them were significant. It was more difficult for NGOs that wanted to provide assistance and set up development projects. In effect, this crucial means of ameliorating the impact of SA and of "King Cotton's" disappearance from the area was unavailable to farmers there.

Organizations of the kind found in San Dionisio and in Santa Rosa resemble what Polanyi (1944) spoke of as a countermovement. According to Polanyi (1944, 131), "the countermovement consisted in checking the action of the market in respect to the factors of production, labor, and land. This was the main function of interventionism." Why was interventionism necessary? Because unchecked market relations would "annihilate" the land and people. Thus, countermovements were primarily self-protective. And those who were most in need of self-protection, the sectors most immediately struck by the negative effect of market relations, were located where support for them was most likely to be found.

Among the "instruments of intervention" that Polanyi (1944, 132) spoke of were protective legislation and restrictive associations (such as guilds and unions). In the context of post-1990 Nicaragua, protective legislation was greatly weakened, as it "interfered with the market" (which it was designed to do). But, associations in the form of unions and producer organizations did spring forth. These associations, such as those found in Esquipulas, San Dionisio, and Santa Rosa, were principally concerned with protecting their members, whether from intermediaries, usurers, or simply from the vicissitudes of the market. And, it would appear that the countermovements that these groups symbolized made a difference with regard to the population's relations with Nicaragua's expanding market. That is, small farmers in the four municipalities where I conducted fieldwork evidenced differing levels of organization, which corresponded to the indicators describing their access to agricultural resources and their economic well-being in the 1990s.

In addition to stressing the importance of the product these farmers specialized in and their organizational levels in explaining how they fared in the context of SA, it must be recognized that both of these factors were related to the opening up of Nicaragua's economy and society to international forces. Although the breaking down of barriers to trade clearly hurt some sectors of Nicaragua's producer population, the country's dairy farmers actually benefited from this change in the 1990s and the first half of the 2000s. At the same time, the linkages between local and internationally based organizations multiplied after 1990 as the latter explicitly sought to develop relations with actors in civil society, in part in response to the shrinkage of the state. The state generally encouraged this development, as it attempted to relinquish many of its former responsibilities.

In sum, under certain circumstances the effect of SA can be modified, if not made to work in favor of small farmers. Occasionally (notably, for farmers in only one of these four municipalities), trade liberalization brought about by it will generate benefits for small farmers. However, given the general pattern of small farmers being food crop producers, unless a new export market opens up for what they have traditionally produced, they will probably be hurt by trade liberalization. Even where small farmers have been induced to switch to export crop production, the net effects of this shift have been quite mixed (see Conroy, Murray, and Rosset 1996; Murray 1997; Kay 1997). Thus, those small farmers benefiting from the opening to international markets will likely remain a minority. Yet at least some of my interviewees who had kept their domestic-market orientation had been able to buffer themselves from the full effects of SA. Here, participation in an organization that could make up for some of the resources they had lost as a consequence of SA's austerity measures was what provided them with such a buffer.

Conclusion

The findings from my survey, and related interviews, indicate that SA had a similar effect on Nicaragua's small farmers to what it had elsewhere in the region.[35] Government austerity measures greatly reduced their access to productive resources, as well as social services. Moreover, trade liberalization and currency devaluations had a deleterious effect on prices for production oriented toward the domestic market. This combination of reduced access to agricultural resources and poor prices for the goods they produced pushed many of these farmers toward mere subsistence production.

The effects of SA in Nicaragua stood out, though, in the sense that it was part and parcel of a larger process of societal change involving a rapid retreat from socialism. Hence, a number of the resources that SA took away from the country's small farmers had been more available to them before 1990

35. See further Rello (1999), Conroy, Murray, and Rosset (1996), Carter and Barham (1996), Gwynne and Kay (1997), Kay (2002), and Edelman (1999).

than they had been for small farmers elsewhere in the region. The agrarian reform that was initiated in 1979 had reached out to assist them in ways not found in most of Latin America. This came to an end in 1990, when Nicaragua's small farmers were once again relegated to a position of low priority for the government.

And, Nicaragua's privatization process was of a piece with both the SA program and the country's overall retreat from socialism. It underscored the new agenda of the government in terms of reasserting the importance of the traditional private sector, with its private determination of how earnings would be spent. The process also generated a degree of insecurity among agrarian reform beneficiaries that, when combined with the other effects of SA, encouraged the parcelization of cooperatives and the sale of their land.[36] Thus, the position of the country's small farmers was further weakened.

The net effect of these various dynamics was eerily similar to what Karl Polanyi (1944) had described as the outcome of an earlier era's foray into liberalism. With the expansion of market relations that the first period of liberalism gave way to, agriculture and its producers suffered disproportionately. The smallest of farmers bore the brunt of that social experiment.

To the extent that some of my informants had held their own, or even experienced economic advancement in the 1990s, having as their principal product a commodity that was doing well in the international market was a key explanatory factor. But, their participation in community and other kinds of organizations that filled some of the gaping holes left by the cutbacks in government resources directed at their production was perhaps even more crucial. Here, too, we find a resonance with Polanyi's (1944) depiction of the first wave of liberalism—the emergence of self-protective mechanisms, what he calls countermovements, in response to the ravages brought about by it.[37] Other factors besides these two might have served to soften the blows of SA for small farmers elsewhere in Latin America.[38] Yet, in the

36. Admittedly, only a minority of those interviewed had been left with no land following land sales in the 1990s (5 percent), though the fact that they turned up in this survey that specifically sought out "practicing" small farmers was noteworthy.

37. Edelman (1999) also found Costa Rica's peasants to have greatly increased their organizing activities in the face of the economic devastation wreaked on them in the 1980s.

38. In Chile, the logic of the government's post-1993 efforts at "reconversion" of peasant production would seem to suggest that its promoters saw their shift to more profitable products (i.e., the pursuit of more overtly capitalist behavior) to offer such potential (Kay 1997, 2002). Drawing on both the Chilean and Guatemalan cases, avoidance of indebtedness through limited credit use might also help to sustain small-scale production (see Murray 1997; Conroy, Murray, and Rosset 1996).

Nicaraguan case, these "ameliorating" factors did not completely override the effect of the new political economic agenda even where they did exist.

More important, those small farmers who benefited from them remained a minority. The majority simply had to reduce their production, as well as their consumption, at the same time as they accelerated their search for options off the farm—an alternative that likely yielded few results. Where minorities of small farmers who produced "just the right crop" or who received the limited resources that NGOs and other organizations had to offer did sprout up, they came to represent those few who were fortunate enough to rise upward (or just to stay afloat) as the process of social differentiation resulting from the country's new political economy moved apace. For the majority who had not gained entrée into this select group, their wait for the benefits of SA and the larger retreat from socialism to "trickle down" continued.

PART 3

The Reconfiguration of Cuban Socialism

5

Cuba's Post-1990 Economic Strategy

Small farmers in Cuba have also been affected by an economic reform that was at the core of that country's social and economic transformation of the 1990s. Rather than engaging in a retreat from socialism, though, the Cuban government sought to reconfigure it in ways that would allow for its essence to remain intact despite the multiple forces that militated against such a political economic orientation—including the country's altered relationship with the former Council of Mutual Economic Assistance (COMECON) countries—in the 1990s. The reconfiguration of socialism brought major changes in policy that introduced market relations into some parts of the economy where they had been absent in the past. Yet the state maintained its central role in planning and in certain areas of production and distribution. Modifications in agricultural policy were key in this process. But, in stark contrast to the marginal role small farmers had been relegated to in Nicaragua's new political economy, Cuba's small farmer sector was seen as containing the potential to revive agricultural production, which was crucial for the country's economic recovery.

As in Nicaragua, Cuba's economic reform of the 1990s was set against the backdrop of an agrarian reform that had been initiated following the overthrow of the previous regime. In the Cuban case, however, state farms (*granjas*) had come to play a much larger role in that country's agricultural sector than they had in Nicaragua: by the late 1980s, 82 percent of Cuba's agricultural land was controlled by state farms, in contrast to 13 percent in Nicaragua.[1] Even though some of Nicaragua's state farms were quite extensive in acreage, given their origins in confiscated *latifundios* (large plantations), the "bigger is better" model of organizing production held greater sway in Cuba. Moreover, the presence of a large wage labor force in Cuban agriculture before 1959 contributed to the decision of government policy makers to further this process of proletarianization of the rural population through employment on state farms. Nicaragua's agrarian reform, in turn reflecting the previous social organization of production, redistributed a significant amount of land to cooperatives and individuals so that they controlled more land than the workers on state farms.

As described in earlier chapters, the privileging of small- (and medium-) sized producers underwent a reversal in Nicaragua in the 1990s, with the retreat from socialism there. In Cuba, too, some very fundamental tenets of earlier policy making were dramatically revised in the 1990s as part of that country's reconfiguring of socialism. Yet these revisions were part of what made Cuba's economic reform so distinctive. That is, some features of its reform resembled aspects of reform imposed elsewhere, including Nicaragua. These features included a partial opening up to the international economy, cuts in some subsidies, a reduction of employment within the state sector, and efforts to expand the government's tax base. But, it also differed in important ways from structural adjustment–style economic reform. Evidence of this could be seen in the priority that was placed on food crop production for the domestic market, in contrast to the agricultural policies pursued by the Nicaraguan government and those of many other countries attempting to stabilize and to adjust their economies.

The modification of priorities toward food crops was part of a much larger process of change that characterized agriculture in Cuba as of 1990. That process had a number of components. These included the downsizing of production through the dismemberment of most state farms, and their reshaping into multiple cooperatives, and the subdivision of cooperative production into work units composed of a few members. It also entailed partial liberalization, as well as incentivization, of the marketing of agricultural products. At the same time, state farms ceded land to tens of thousands of people, so that they could engage in individual subsistence production. The state invested in the development of organic inputs for domestically oriented

---

1. The figure for Cuba is from Valdés Paz (1997, 147) and for Nicaragua from Dirección General de Reforma Agraria, MIDINRA (unpublished data, 1988).

agriculture. In most of these senses Cuban agricultural policies were distinguished from policies that usually accompany economic reform.

Given both the similarities and the differences between Cuba's economic reform and those carried out elsewhere, the question of the former's effect on small farmers and their production seems most relevant. My exploration of this issue will begin with an overview of Cuba's process of political economic change in the 1990s. Thus, it will become clear that the government's initial reaction to the crisis was to focus on modifying the country's foreign economic relations, but this was altered over time to include the domestic economic and productive structure as well as the state's own fiscal and other policies toward it. The chapter will also describe the larger shift in agricultural policy, with all of its component parts, that coincided with the reform. Finally, the social impact of the reform will be addressed. That impact followed Szelenyi and Kostello's (1996) predictions about the social implications of adopting a socialist mixed economy, as well as Polanyi's (1944) more general thesis that inequality will increase with an expansion in market relations.

Yet I will also illustrate the extent to which the agricultural sector was bolstered through the reform process, thereby distinguishing the Cuban reform from others. This was accomplished through the provision of new resources to its producers and through their privileged position within Cuba's reconfigured model of socialism. With this backdrop sketched out, I will then go on—in the following chapter—to look specifically at what these changes meant for the country's small farmers.

Cuba's Economic Crisis and Reform

Cuba's economic reform brought about changes in the structure of the economy that were more thoroughgoing than any that had been imposed since the initial stages of the revolution. These changes included major modifications in the agricultural sector but were also of a more global nature and affected many aspects of the economy. Essentially, they constituted a reconfiguration of Cuban socialism.

The reform was set in motion in response to the profound economic crisis that gripped the country in the early 1990s. The depth of the crisis Cuba faced was acknowledged by the government when it forewarned its population in 1990 that the country's economy and society would be confronting the equivalent of wartime conditions, what it called "The Special Period in Peacetime," for the foreseeable future. The year 1993 is now seen as the turning point in Cuba's economic crisis, after which the economy began to recover from the free fall that characterized it during the preceding few years. Between 1990 and 1993, the country's gross domestic product (GDP) fell by 35 percent (or 50.2 percent between 1989 and 1993).[2] Moreover, exports dropped by 80 percent between 1990 and 1993 and imports by roughly 75 percent (Monreal 1999, 23; and Estay R. 1997, 21). Clearly, a crisis of these proportions would require drastic measures to bring about its reversal.

The origins of the crisis can be found in the structural imbalances that were already beginning to cause difficulties by the latter part of the 1980s. These imbalances stemmed from the country's continuing dependence on agro-exports (80 percent of its export earnings came from sugar in 1989) and imports (especially of fuel oil, food products, and inputs), falling productivity and underutilization of the labor force, overlong investment cycles, and the growing lack of effectiveness in the use of basic funds.[3] However, the disso-lution of the COMECON accentuated, in an extreme fashion, all of these weaknesses by eliminating the entire logic on which the economy had rested since the mid-1960s. Production in Cuba had, to a large extent, been planned according to the role the country played within the COMECON. Exports and imports were disproportionately dependent on relations with the COMECON's members. The drop in exports and imports reflects the degree of trade dependence Cuba had with the COMECON, and within the COMECON with the USSR in particular.[4] Furthermore, Cuba had received preferential prices for some of its key exports—especially sugar—within the COMECON marketplace, which had served as a buffer from the winds of the international economy (cf. Pastor and Zimbalist 1995; Zimbalist 1992). The COMECON's disintegration threw these trade relations into disarray and quickly led to a massive drop-off of them.

In addition, shortly after the crisis began, the U.S. government tightened the embargo it had in place against Cuba for the previous three decades. The Torricelli Bill, which was voted into law in 1992, among other things brought to an end the possibility of U.S. subsidiaries doing business with

2. For the former figure, see Estay R. (1997, 18), CEPAL (1997a, 61), and Monreal (1999, 23); for the latter figure, see Pastor and Zimbalist (1995, 707).

3. Estay R. (1997) draws on the work of many authors in assembling this list; see also, Pastor and Zimbalist (1995), Carranza Valdés (1992), and Eckstein (1997).

4. Approximately 70 percent of imports and exports stemmed from trade with the USSR (Álvarez González 1995, 5).

Cuba. More than one hundred subsidiaries had engaged in $718 million in trade with Cuba in 1991 (Estay R. 1997, 24). The tightening of the embargo further complicated an already problematic economic situation.[5]

Initially, the reactions of Cuban policy makers were geared almost exclusively toward modifying the country's foreign economic relations. They sought to develop new trade relations and products for trade, as well as to bring in foreign investment. Gradually, Cuba succeeded in making strong strides toward all of these objectives. Given the limited prospects for Cuba's traditional exports to pull the economy out of the tailspin that it was in, new products were proposed to do this: tourism and pharmaceuticals. Although the pharmaceutical sector did not live up to their expectations in the early 1990s,[6] tourism did take off: by 1994, income from tourism surpassed Cuba's previous primary foreign exchange earner, sugar, and the number of tourists and gross earnings from tourism grew on average annually by 24 percent and 44 percent, respectively, between 1992 and 1998, generating almost $1.8 billion (U.S.) in gross earnings in 1998 (CEPAL 1997a, 167; 2000, A.107; and CEPAL/INIE/PNUD 2004, 18). (See Table 5.1.) This upward trajectory continued through the rest of the decade, dipping somewhat in 2001 and 2002 (as was true of tourism worldwide after 9/11), before climbing again between 2003 and 2005 (CEPAL/INIE/PNUD 2004, 18).

Moreover, tourism was at the forefront of the areas that foreign investment was attracted to. Foreign investment within tourism took the form of joint ventures with the Cuban state to construct new hotels and contracts to manage and upgrade existing ones. Yet this was simply the beginning of a flow of foreign capital that also reached into nickel mining, construction, real estate, energy generation, telecommunications, agricultural production, and marketing. Joint ventures with foreign capital went from 20 in 1990, to 80 by 1992, 226 by 1995, and 340 by 1998.[7] This growth continued through 2002 before dropping off somewhat afterward (see Pérez Villanueva 2004b, 172). Capital flows and income from tourism helped to offset the reduced earnings from Cuba's traditional exports (to be discussed later).

5. Then, in 1996, the Helms-Burton bill was passed into law. This bill further tightened the embargo and made it congressionally mandated. Among other things, it punished foreign companies that "trafficked" in Cuban properties that had been confiscated from U.S. citizens or Cuban nationals who later became U.S. citizens (cf. Gunn Glissold 1997).

6. See further Monreal (1999) and Pastor and Zimbalist (1995).

7. CEPAL (2000, table A42) sets forth significantly lower total figures for each year but documents the same overall trend.

Table 5.1   Foreign exchange earnings from tourism (millions of U.S. dollars[a])

	1990	1991	1992	1993	1994	1995	1996
Gross earnings	243	402	550	720	850	1,100	1,333

SOURCES: Data for 1990–97: CEPAL 2000, table A.107; data for 1998–2000: CEPAL/INIE/
PNUD 2004, table I-15; data for 2001–5: ONE 2007, table 15.11.
[a]Data for 1990–2000 are in U.S. dollars, and data for 2001–5 are in Cuban Convertible Pesos
(CUC). US$1 = 1 CUC during this period.

Even though progress was being made on all of these fronts, by mid-1993, the focus of government efforts to deal with the crisis had to be broadened beyond foreign economic relations to include reform of some aspects of the domestic economy and productive structure. The first indication of a broadening of the reform's focus was the legalization of foreign currency holdings by Cubans in August 1993. Previously, it was illegal for Cubans to have and use any foreign currency. This measure sought to undercut the black market in U.S. dollars, and the purchase of items with them, as well as to increase the state's foreign exchange earnings through a major expansion in the state-owned retail store system for mostly imported goods sold in U.S. dollars.[8] Later, in 1995, state-owned exchange offices offered the government another means of absorbing U.S. dollars the population had access to through foreign firms, tourists, and family remittances.

These sources of funding had come to provide the population with steadily greater, even noteworthy, sums of foreign currency. This was especially true of remittances. Whereas only in-kind remittances were legally permitted before the August 1993 measure, by the late 1990s, the role of monetary remittances in the economy was similar to that found elsewhere in Latin America. Official figures show an increase in them of 45 percent between 1994 and 1995 (with their percentage of the GDP increasing from 1.4 percent to 2.7 percent during that same period; calculated from Barberia 2004, 368). Their growth leveled out toward the latter 1990s at approximately 300 percent more in absolute amounts than in 1994, reaching $740 million (in U.S. dollars) in 1999.[9] The remittances facilitated an improved standard of living

8. In fact, sales in U.S. dollars had come to represent 54 percent of all domestic retail sales by early 1999 (Monreal 1999, 26).

See, also, n. 3 in Chapter 1. After this point in time, "convertible pesos" were the currency used in these stores. Cubans earning regular pesos could purchase the "convertibles," but it was a costly endeavor for them as they were worth substantially more than regular pesos (they exchanged at roughly twenty pesos for one "convertible peso").

9. Mesa-Lago (2003, 83) quotes other Cuban economists who calculate that by 2002 this figure was between $800 million and $1.1 billion (U.S.).

1997	1998	1999	2000	2001	2002	2003	2004	2005
1,543	1,759	1,901	1,948	1,840	1,769	1,999	2,114	2,399

for their recipients, while expanding foreign currency generation for the government.

The legalization of foreign currency holdings was shortly followed by an even more far-reaching measure: the creation of Basic Units of Cooperative Production (UBPCs) through the subdivision of some of the existing state farms. The downsizing of agricultural production was, thus, effectively initiated. As described at length elsewhere (Rosset 1997; Pérez Marín and Muñoz Baños 1992; Enríquez 2000), the predominant model of agriculture in Cuba before 1990 was large-scale farms, as typified in the country's state farms. To a lesser extent, it was also true of the Agricultural Production Cooperatives (CPAs), which were collectively owned and operated cooperatives. With the drastic drop in input imports in the early 1990s came a realization that this "bigger is better" orientation would have to be modified because material conditions could no longer sustain it. At the same time, government policy makers acknowledged that production levels were much higher on smaller production units. Even though a number of studies had pointed to this tendency in the 1970s and 1980s (see Forster 1982; Benjamin, Collins, and Scott 1984; and Lehmann 1985), it was the economic crisis of the early 1990s, and especially the dismal sugar harvest of 1992/93 (see Table 5.2), that led to the historic decision to subdivide the state farms into UBPCs (i.e., cooperatives). The principal objectives behind this measure were to improve the level of efficiency and raise productivity in agriculture.[10]

The UBPCs were something of a hybrid. In some senses they functioned like CPAs, with their members working the land collectively. Together, their members owned the equipment and the fruits of their labor. But, in contrast to CPAs, the land they worked continued to be state property, only given over to them in usufruct. The UBPCs had a small administrative staff and

---

10. For further discussion of the reasons behind the formation of the UBPCs, see Figueroa Albelo (1996) and Pérez Rojas and Torres Vila (1996).

Table 5.2   Harvested area (hectares) and production (millions of metric tons) of sugarcane, 1990/91–2005/6: Overall and by sector

	1990/91	1991/92	1992/93	1993/94	1994/95	1995/96	1996/97
Area							
State sector	1,219,200	1,230,600	1,013,200	63,800	60,100	97,100	111,300
Nonstate	233,000	221,100	198,500	1,185,100	1,117,300	1,147,400	1,135,000
Total	1,452,200	1,451,700	1,211,700	1,248,900	1,177,400	1,244,500	1,246,300
Production							
State sector	64.9	54.4	35.4	2.2	1.7	2.7	2.7
Nonstate	14.8	11.9	8.3	41.0	31.9	38.6	36.2
Total	79.7	66.3	43.7	43.2	33.6	41.3	38.9

Sources: Data for 1990/91–2001/2: ONE 2004, table X.6; data for 2002/3–2005/6: ONE 2007, table 9.4.

administrator, all of whom were ratified by the membership. They were nominally independent from the larger state enterprise that they formerly composed a piece of, but they continued to be closely tied to it for determining production schedules and obtaining services.[11]

Although the UBPCs' formation was initially planned to be a gradual process, they increased in number at a very rapid rate. By the end of 1996, there were 1,656 UBPCs (Cárdenas Toledo 1998, 166).[12] They played their largest role in sugarcane production, where they represented 73 percent of its acreage, but were also important in citrus, rice, and livestock production (calculated from CEPAL 2000, A.73). With the spread of UBPCs during this period, the size of the state farm sector dropped from representing 82 percent of Cuba's agricultural land to 32.7 percent (Valdés Paz 1997, 147; CEPAL 2000, 313, respectively). Because of the size of the UBPC sector, its establishment had a great effect on agricultural output. The performance of the UBPCs during the first several years of their existence, however, fell significantly short of expectations for them. A variety of factors caused this, including the extremely difficult macroeconomic context in which they were formed, which was characterized by tremendous input shortages; the technological and organizational changes their establishment required, which implied "growing pains"; problems with labor productivity, which affected yields; and the continuing overbearance of the state farms out of which they had sprung (see Nova González 2003; Rodríguez Eduarte 1997). Nonethe-

11. The Colectivo de Autores (1996), Bu Wong et al. (1996), and Enríquez (2000) provide further details on the functioning of the UBPC sector.

12. By the end of the decade, this figure had dropped to 1,610 through the merger of some UBPCs (Nova González 2003, 9).

1997/98	1998/99	1999/2000	2000/2001	2001/2	2002/3	2003/4	2004/5	2005/6
67,500	82,300	89,700	86,600	89,600	45,200	44,200	21,800	13,800
98,800	913,500	951,200	920,500	951,600	598,600	616,800	495,400	383,300
1,048,500	995,800	1,040,900	1,007,100	1,041,200	643,800	661,000	517,200	397,100
1.7	2.6	2.9	2.6	2.8	0.6	1.3	0.4	0.4
31.1	31.4	33.5	29.5	31.9	21.5	22.5	11.2	10.7
32.8	34.0	36.4	32.1	34.7	22.1	23.8	11.6	11.1

less, through consolidation of weaker UBPCs with stronger ones and price increases, profit levels rose during the 1990s. Thus, between 1996 and 2001, the percentage of profitable UBPCs grew from approximately thirty[13] to approximately seventy.[14] This improved performance suggested that, with time, downsizing would bear fruit.

The new emphasis on downsizing that lay behind the dismemberment of much of the state farm sector was also promoted in the CPA sector of agriculture. A novel strategy of organizing work was being experimented with on the CPAs, as well as the UBPCs, known as "linking the person with the area." This form of organization represented an informal subdivision of land into relatively small units that would be worked by a fixed group composed of a few cooperative members. These members were to be rewarded for their efforts to ensure high levels of productivity by receiving a portion of the value of "excess production" (i.e., what was left after the quotas of the state purchasing and distributional agency—Acopio—had been met). The idea was to approximate organization and incentives on small private farms. "Linking" was begun on an experimental basis in 1993/94 but had extended its reach notably by the late 1990s. Agricultural policy makers expressed some concern about letting this process go too far (e.g., *Granma* 1998). In essence, they sought to achieve the yields of the small farm sector, while ensuring that cooperatives remained a key form of organizing small farmers.

13. Rodríguez Eduarte (1997). See also Paneque Brizuelas (1997) and Nova González (2003). And CEPAL (2000, table A.13) shows diminishing financial support from the state to the UBPC sector in the latter 1990s.

14. Pagés (2002) and Nova González (2003, 9). Profitability levels varied notably by product (see Nova González 2004b).

Yet smaller-scale agriculture in an individual fashion also blossomed in the early 1990s, with the emergence of the *parcelero* sector of agricultural producers. Beginning at least as early as 1992, unused state farmland was turned over in usufruct to those who wanted to farm it. These *parceleros* were given a usufruct title for the land so that they would have the security that they would be able to work the same plot year after year.

The initial idea was that production on these *parcelas* would largely be of a subsistence nature. To try to expand export crop production, though, *parcelas* were also made available for people who were willing to engage in tobacco production. Coffee *parceleros* were granted usufruct rights as well. The size of the plot allotted varied according to the number of family members working on it. The average tobacco *parcela* was 3.4 hectares, the average coffee *parcela* was 10.9 hectares, and the average food crop *parcela* was .25 hectares (CEPAL 2000, 315–16). By fall 1998, nearly 75,000 *parceleros* had received 128,000 hectares. Thus, the nonstate sector of agriculture underwent a remarkable expansion with the creation of the UBPC and *parcelero* sectors. These changes, but especially the subdivision of state farms into UBPCs, have together been referred to as Cuba's third agrarian reform.[15]

Industrial production, however, was also affected by the economic crisis and reform. At first, because of limitations of imported inputs, Cuba's industrial sector experienced a serious slowdown. Then, because of its fundamentally domestic orientation and its high content of imported inputs, it became a less prioritized sector of the economy. These changes were evident in the 37.5 percent drop in the manufacturing industry's GDP between 1989 and 1993 (calculated from CEPAL 2000, A.10). Moreover, to reduce subsidies to nonprofitable state enterprises and to force greater fiscal discipline, a partial closing down of industries producing goods for the Cuban population occurred. This led to a "rationalization" of the labor force, causing a reduction in overall public sector employment between 1990 and 1995 of 745,000 workers, with the biggest drop occurring between 1993 and 1994 (calculated from CEPAL 2000, table A.46). At the same time, throughout the state, a drive was under way to achieve greater competitiveness and to move toward the modernization of those enterprises that remained in existence.

The cuts in subsidies to industry and streamlining of employment rolls within this sector were complemented by a reorganization of the state's cen-

15. The first was initiated in 1959, and the second in 1963. See further, Valdés Paz (1997) and Burchardt (2000).

tral administrative structure. The number of agencies and ministries within this structure were reduced, as was the size of its workforce, which dropped from 19,800 to 8,000 between 1994 and 1995 (CEA 1996, as cited in Estay R. 1997, 34). The goal was to adjust the state structure to the new exigencies.

In addition to decreasing public sector employment, given the extremely inflated prices of the black market and the limited array of goods available through the state rationing system, by 1993, the peso salaries of those who still had employment did not provide them with the means to purchase much. It was within this context that the state undertook several extra reform measures. First, in September 1993, the government legalized a restricted number of forms of self-employment, which increased periodically after that time. Several objectives were behind this initiative: to allow for other forms of employment now that the state and its enterprises could no longer ensure full employment; to legalize forms of employment that were already in existence; and to facilitate the availability of certain services that the state was no longer in a position to provide to the extent for which demand existed. The range of self-employment options initially legalized included providers of various forms of transportation, repair people, barbers, cooks, and carpenters. According to official figures, the population employed in this sector went from zero to 121,100 in the year following its legalization (CEPAL 2000, table A.48). Although growing further in 1995, the size of this sector stabilized around 120,000 during the remainder of the decade. This represented approximately 3.3 percent of the labor force (CEPAL 2000, table A.48).[16]

Slightly over a year later, legislation was enacted that allowed for the establishment of farmers' markets, known as Mercados Agropecuarios.[17] With the announcement of this decree, it once again became legal, as it had been in the 1980–86 period, for farmers to sell their excess production in markets in which prices were established by the laws of supply and demand. Once they had met Acopio's quotas, farmers were free to sell the remainder of their production at a higher price in the newly established markets. Despite being taxed on their earnings, prices for the produce sold in the markets were significantly higher than those offered by Acopio.[18]

16. CEPAL/INIE/PNUD (2004, A.22, 177, respectively) show this percentage vacillating between 3.0 and 4.1 from 1997 and 2001, and state that 153,000 people were employed in this sector in January 2003.

17. For further information concerning the Mercados Agropecuarios, see Torres Vila and Pérez Rojas (1996b), Nova González (1995), and Enríquez (2000).

18. See further Enríquez (2000, n. 64); and ANEC et al. (2000).

Yet, the 1994 market liberalization differed from that of the 1980s in an important sense. In the 1980s the state sector (which controlled the vast majority of Cuba's agricultural land then, but whose weight in production varied somewhat by crop) was not permitted to participate in the markets. Nevertheless, in the 1990s state farms, non-sugarcane UBPCs,[19] CPAs, military farms, and independent farmers were allowed to sell there. In essence, the 1994 legislation contained within it the potential to affect a much larger portion of marketable produce in this second experiment with liberalization.

The markets' opening in 1994 represented a recognition of the strength of the black market for produce and a desire to undercut it by making agricultural produce more readily accessible to the population through legal channels. Likewise, it was assumed that the option to market produce in the Mercados would incentivize production expansions, which would also help to bring prices down, to say nothing of facilitating the overall effort to improve the availability of food products for the population. Finally, as part of another effort under way—to reduce the extremely high level of liquidity in the economy—the Mercados absorbed cash that the population previously had been unable to spend because of the lack of supply. The excess cash had contributed to the growing inflation that characterized the economy.

By the end of the 1990s, the Mercados had met a number of these objectives, while falling short on others. Sales in the markets grew steadily over the course of the 1990s (ANEC et al. 2000), translating into a real increase in the availability of produce for consumers. Prices, however, after falling initially (especially in relation to the black-market prices that had existed before the Mercados' opening), then became stagnant at relatively high levels before rising again at the end of the 1990s (ANEC et al. 2000; Nova Gonzalez 1998). Despite the growth of the role of state farms in the market, which had been assumed to carry enough weight to force prices down, production expansions were not sufficient to seriously affect prices (production levels will be addressed later). Although liquidity levels dropped notably in the first few years after the Mercados were established (by 21 percent between 1993 and 1994 and another 10 percent between 1994 and 1995; calculated from CEPAL 2000, table A.26), they rose again in the latter 1990s.[20]

Nonetheless, the government also put in place more direct mechanisms

19. The exclusion of the sugarcane-producing UBPCs was designed to avoid a conversion in production to more profitable crops.

20. See CEPAL (2000, table A.26) and CEPAL/INIE/PNUD (2004, A.25) on the continuing upward trend in liquidity levels.

to encourage the expansion of production during the 1990s, in the form of price incentives for certain agricultural goods.[21] The list of incentivized goods was weighted heavily in the direction of export products (e.g., coffee, honey, tobacco, and citrus crops), as they generated foreign exchange earnings for the country.

The incentive systems applicable for each product varied somewhat. For example, coffee and citrus producers received partial payment in credit at one of the government-run stores for imported goods where everything was sold in U.S. dollars. In contrast, because of a campaign during the latter part of the 1990s to expand tobacco cultivation, incentives for it were given in cash in U.S. dollars so that tobacco farmers could use the money how they saw fit. Moreover, incentives for its cultivation began with its planting (assuming that it was planted according to technical specifications). Then, in addition to the official price farmers received for their crop, they were given a bonus per hundredweight in U.S. dollars.

Have farmers responded to the multiple incentives they have been presented with by expanding their production? The picture is not entirely clear. Tables 5.3 and 5.4 detail the pattern of acreage under cultivation and production levels of Cuba's key crops—excluding sugarcane—during the 1990s, with Part 1 of each showing overall patterns and Part 2 distinguishing by producer sector. (Table 5.2 shows all of these data for sugarcane.) In analyzing these tables, it is essential to bear in mind several things: that the country's economy did not begin to turn around from its free fall until 1994 and then did so only gradually, which affected its ability to import many of the inputs required in agricultural production (see further Chapter 6); that the formerly dominant state farm sector began to be downsized in late 1993, thus it could be expected that some time would be needed for the new UBPC sector to become fully functional, but a shift in the weight of the state versus nonstate sector in production should also become apparent in the few years following this point in time; and that the Mercados were first opened in late 1994 and other incentive systems were put in place around the same time.

21. Abbasi (1997) posits that incentives for food crop production were also offered to the small-farm sector before 1990, in the aftermath of the closing of the first farmers' markets in 1986. That is, because food crop production has always been crucial—and the role of small farmers in it likewise—the state has consistently, over time, been forced to recognize their importance through more favorable prices and the provision of inputs. Yet, I would argue that recognition of the importance of small farmers in food production extended way beyond favorable prices and input provision in the 1990s, as seen through, among other things, the massive downsizing of agricultural production and the turning over of land to individual farmers who were willing to cultivate it.

Table 5.3.1 Overall harvested area for Cuba's principal crops, 1990–2005 (hectares)

Products	1990	1991	1992	1993	1994	1995	1996
Root crops	152,313	155,794	170,761	150,865	146,961	137,818	142,954
Plantains/bananas	99,500	108,200	123,900	123,800	132,000	122,200	116,400
In development	44,197	57,834	62,527	61,672	60,810	48,265	42,065
Already producing	55,303	50,366	61,373	62,128	71,190	73,935	74,335
Vegetables	125,671	117,927	109,402	91,714	85,676	76,489	86,091
Corn	74,604	68,957	71,447	68,188	77,644	76,891	89,227
Beans	45,227	45,650	47,675	49,285	55,315	44,084	46,861
Citrus	144,900	141,400	134,000	127,300	118,700	115,900	86,000
In development	30,234	27,434	19,841	14,423	16,058	14,453	6,625
Already producing	114,666	113,966	114,159	112,877	102,642	101,447	79,375
Other fruits	72,100	69,500	66,400	57,700	60,000	51,900	54,600
In development	20,033	22,856	18,650	13,018	14,345	13,873	13,362
Already producing	52,067	46,644	47,750	44,682	45,655	38,027	41,238

SOURCE: Calculated from ONE 2007, tables 9.5, 9.6.
[a]Includes area of nonspecialized producers and home gardens. Before this time, these areas were not included.

With these provisos in mind, it is evident that the acreage and production levels of certain crops dropped off notably during the decade of the 1990s. Others first experienced a decline before rebounding after the mid-1990s, and others simply climbed steadily upward during this period. In disaggregating the data by crop, a bit more light can be shed on these dynamics. A major push to expand plantain/banana production occurred in the early 1990s, as part of Cuba's newly inaugurated national Food Program (see Enríquez 1994). The expansion was heavily technology dependent, as seen in its use of more sophisticated types of irrigation systems, which were not well suited for the conditions of the Special Period. Yet the acreage in plantain/ bananas still grew steadily through the mid-1990s, after which it leveled off before experiencing more growth in the mid-2000s (see Table 5.3). Production levels moved in a similar fashion, although achieving less stability (see Table 5.4). In contrast, after initially increasing, acreage in root crops decreased in the mid-1990s before experiencing an erratic pattern of increases after that time (see Table 5.3), most especially in the state farm sector (even taking into account the roughly 40,000 hectares of root crops that shifted from it to the nonstate sector in the 1993–94 period[22]). However, production levels for root crops tended to be more favorable than might have been expected (see Table 5.4), even as acreage fell. While vegetable acreage first

22. This is, admittedly, a rough estimation by the author based on the simultaneous dropoff that occurred in the state sector and expansion in the nonstate sector between 1993 and 1994.

1997[a]	1998[a]	1999[a]	2000[a]	2001[a]	2002[a]	2003[a]	2004[a]	2005[a]
193,624	205,000	165,071	167,711	190,519	199,669	205,650	241,100	251,648
121,800	120,000	118,400	126,200	123,500	125,000	144,400	143,000	137,100
19,170	32,604	23,988	14,610	12,172	16,544	21,169	19,979	41,709
102,630	87,396	94,412	111,590	111,328	108,456	123,231	123,021	95,391
104,782	139,434	143,635	179,882	201,745	232,532	265,403	312,549	311,732
159,016	135,444	118,282	126,313	128,253	122,240	132,247	145,209	155,580
106,065	103,240	92,450	105,722	103,249	100,227	106,779	112,201	94,821
80,700	76,200	69,100	66,600	65,200	65,200	65,200	66,100	60,800
2,021	207	<5,286>	435	373	<6,137>	<3,852>	7,939	4,552
78,679	75,993	74,386	66,165	64,827	71,337	69,052	58,161	56,248
56,100	56,900	54,600	61,900	63,900	63,800	78,200	83,200	83,800
<1,822>	<16,462>	<9,677>	<22,150>	<21,961>	<23,027>	<14,366>	6,634	2,791
57,922	73,362	64,277	84,050	85,861	86,827	92,566	76,566	81,009

underwent a precipitous decline between 1990 and 1995, it then initiated a slow ascent that continued through the rest of the 1990s and into the next decade (see Table 5.3). Production levels of vegetables did not experience a downturn until the mid-1990s and rose again in the late 1990s (see Table 5.4). This latter pattern continued into the next decade. This would indicate that farmers were somewhat more successful at maintaining yields than acreage.

Each of these types of crops is sold in the Mercados Agropecuarios, thereby providing their producers with that particular incentive. Nonetheless, they did not benefit from the same price incentive systems of the specific products mentioned earlier. Additional factors undoubtedly also came into play. In an analysis of the multiple factors that affected agricultural production and, therefore, the supply of goods available to consumers, Nova González (1995) lists: resource limitations (here inputs shortages were key); the prohibition against certain goods being sold in the Mercados and the loss of that incentive; the exclusion of the sugarcane UBPCs from participation in the Mercados, as they contained the potential to increase supplies there; insufficient autonomy (especially in terms of production planning) of the UBPCs from the state sector; the state's strong role as produce purchaser, rather than regulator; insufficient producer incentives; inefficiencies in resource use by the state and UBPC sectors; and the very high profit margins obtained by private farmers who channeled their products through

Table 5.3.2 Harvested area by sector for Cuba's principal crops, 1990–2005 (hectares)

	1990	1991	1992	1993	1994	1995	1996
State sector							
Root crops	92,231	102,997	112,657	94,908	52,354	41,513	43,665
Plantains/bananas	58,800	68,000	83,000	83,100	51,300	43,500	41,900
In development	33,715	37,013	44,133	40,672	26,406	18,152	14,794
Already producing	25,085	30,987	38,867	42,428	24,894	25,348	27,106
Vegetables	66,042	67,517	62,283	49,792	27,738	22,154	29,214
Corn	37,824	34,502	35,465	31,454	19,871	18,819	22,603
Beans	27,150	28,416	31,590	32,872	19,969	12,881	12,820
Citrus	129,200	125,900	120,400	113,800	61,900	60,500	34,900
In development	26,304	21,588	17,644	12,242	7,659	6,737	957
Already producing	102,896	104,312	102,756	101,558	54,241	53,763	33,943
Other fruits	46,500	45,000	43,000	38,800	21,000	19,100	17,100
In development	13,251	13,764	11,658	9,218	5,800	6,494	4,438
Already producing	33,249	31,236	31,342	29,582	15,200	12,606	12,662
	1990	1991	1992	1993	1994	1995	1996
Nonstate sector							
Root crops	60,081	52,798	58,104	55,957	94,607	96,305	99,289
Plantains/bananas	40,700	40,200	40,900	40,700	80,700	78,700	74,500
In development	10,482	20,822	18,394	20,999	34,404	30,113	27,271
Already producing	30,218	19,378	22,506	19,701	46,296	48,587	47,229
Vegetables	59,629	50,410	47,119	41,922	57,938	54,335	56,877
Corn	36,780	34,455	35,982	36,734	57,773	58,072	66,624
Beans	18,077	17,235	16,085	16,414	35,346	31,203	34,041
Citrus	15,700	15,500	13,600	13,500	56,800	55,400	51,100
In development	3,930	5,846	2,197	2,181	8,399	7,716	5,668
Already producing	11,770	9,654	11,403	11,319	48,401	47,684	45,432
Other fruits	25,600	24,500	23,400	18,900	39,000	32,800	37,500
In development	6,782	9,092	6,992	3,800	8,545	7,378	8,923
Already producing	18,818	15,408	16,408	15,100	30,455	25,422	28,577

SOURCE: Calculated from ONE 2007, tables 9.5, 9.8.
[a]Includes area of nonspecialized producers and home gardens. Before this time, these areas were not included.

the Mercados. According to Nova González (1995), this last factor worked as a disincentive for expanding production because if production levels were to increase, farmers would be required to expend more energy, prices would be reduced, and, consequently, their profits would be too.[23] Cuba also experi-

23. Mesa-Lago (2003) concurs that this last factor represented a clear disincentive for expanding production.
   While Nova González (1995) argues that the effective monopoly the "private sector" had in the marketing of goods in the Mercados represented a block on production expansions as a result of the high profit margins they obtained there, Pérez Villanueva (2004, 64) credits the "buoyancy of the sales of the agricultural market" for increased crop production, "particularly in the nonstate sector."

1997	1998	1999	2000	2001	2002	2003	2004	2005
41,152	35,242	31,699	26,597	30,135	26,879	29,250	31,779	31,866
45,500	44,200	36,800	37,400	32,700	31,200	35,400	32,600	34,000
19,510	17,213	11,898	16,095	12,638	12,158	15,430	14,325	18,183
25,990	26,987	24,902	21,305	20,062	19,042	19,970	18,275	15,817
23,992	24,408	30,014	29,580	31,157	33,900	37,222	43,205	39,005
23,961	19,666	21,684	17,229	17,014	13,684	15,060	14,959	13,624
12,713	11,360	9,414	11,870	10,680	10,092	9,406	8,865	6,975
31,400	31,600	27,900	26,800	26,900	28,400	29,900	28,300	28,800
581	1,911	1,548	2,208	1,073	1,046	6,239	5,826	6,606
30,819	29,689	26,352	24,592	25,827	27,354	23,661	22,474	22,194
19,100	19,200	18,400	20,900	18,300	18,000	23,800	25,700	25,100
5,889	6,436	5,887	10,490	7,581	6,784	10,854	13,088	15,492
13,211	12,764	12,513	10,410	10,719	11,216	12,946	12,612	9,608

1997[a]	1998[a]	1999[a]	2000[a]	2001[a]	2002[a]	2003[a]	2004[a]	2005[a]
152,472	169,758	133,372	141,114	160,384	172,790	176,400	209,321	219,782
76,300	75,800	81,600	88,800	90,800	93,800	109,000	110,400	103,100
< 340>	15,391	12,090	<1,485>	< 466>	4,386	5,739	5,654	23,526
76,640	60,409	69,510	90,285	91,266	89,414	103,261	104,746	79,574
80,790	115,026	113,621	150,302	170,588	198,632	228,181	269,344	272,727
135,055	115,778	96,598	109,084	111,239	108,556	117,187	130,250	141,956
93,352	91,880	83,036	93,852	92,569	90,135	97,373	103,336	87,846
49,300	44,600	41,200	39,800	38,300	36,800	35,300	37,800	32,000
1,440	<1,704>	<6,834>	<1,773>	< 700>	<7,183>	<10,091>	2,113	<2,054>
47,860	46,304	48,034	41,573	39,000	43,983	45,391	35,687	34,054
37,000	37,700	36,200	41,000	45,600	45,800	54,400	57,500	58,800
<7,711>	<22,898>	<15,564>	<32,640>	<29,542>	<29,811>	<25,220>	<6,454>	<12,601>
44,711	60,598	51,764	73,640	75,142	75,611	79,620	63,954	71,401

enced various adverse climatological conditions and pest problems in the 1990s, including a major infestation of the pest thrips in the mid-1990s, which seriously affected the yields of a number of important crops such as potatoes; a severe drought affected the eastern part of the island for most of the second half of the 1990s, and several major hurricanes occurred. Even though it is difficult to determine which factors were most crucial in defining the production patterns we see in these tables, the result was less than overwhelmingly positive.

Shortly after the liberalization of agricultural markets began, markets for

Table 5.4.1 Overall production levels for Cuba's principal agricultural goods, 1990–2005 (tons)

Product	1990	1991	1992	1993	1994	1995	1996
Root crops	702,290	690,446	753,942	568,727	484,537	624,195	742,291
Plantains/bananas	324,204	357,099	514,625	400,018	360,679	399,989	539,426
Vegetables	484,213	490,795	513,679	392,883	322,164	402,281	493,577
Corn	65,045	55,349	58,469	49,449	73,623	80,990	104,325
Beans	12,024	11,847	9,727	8,819	10,771	11,474	14,049
Citrus	1,015,873	826,001	786,980	644,466	504,991	563,539	662,201
Other fruits	218,968	257,632	127,422	68,345	89,110	112,290	102,554
Milk	1,034,400	820,300	622,300	585,600	635,600	638,500	640,000

SOURCE: ONE 2007, tables 9.9, 9.17.
ªIncludes production of nonspecialized farms and home gardens.

artisan and industrial goods were also established. These were supposed to function along the same lines and with the same logic as the agricultural markets. Both types of markets were located primarily in urban areas.

The legalization of self-employment, the opening of the Mercados Agropecuarios, and the artisan/industrial goods markets, brought with them the implementation of a new tax system in 1994. Before this time, income taxes had been implicit within the salary system. However, with the birth of income generation outside the bounds of state control came the imposition of direct taxes on earnings. In fact, taxes collected from the "self-employed" increased by 150 percent between 1994 and 1997, as this sector grew in size (calculated from CEPAL 2000, table A.20). Taxes collected from small-scale agricultural producers grew by 175 percent during this same period. Yet the reach of the tax system extended beyond those who had earnings generated through these specific marketplaces. The new tax laws stipulated that all forms of individual income were supposed to be taxed.

But it was not just individuals who were subject to the new tax system. Although initially its implementation was only partial in nature, eventually all profitable enterprises—state, joint ownership, and Sociedades Anónimas—were to be subject to taxation. In addition, the collection of social security taxes was stipulated under the new tax system. All firms were required to pay a tax for the employment of wage labor. Even though the tax system was still in the process of formation, through it the state had already begun to absorb excess liquidity and to move toward balancing its budget.

As part of this dual effort, prices on some goods and services were increased. Most of the goods affected were categorized as nonessential items,

1997[a]	1998[a]	1999[a]	2000[a]	2001[a]	2002[a]	2003[a]	2004[a]	2005[a]
827,282	817,570	1,059,247	1,230,841	1,380,559	1,437,029	1,843,600	1,946,400	1,801,800
529,300	566,100	603,200	844,900	968,000	729,900	1,112,600	1,215,600	773,500
601,021	846,809	1,442,506	2,372,673	2,676,475	3,344,710	3,931,200	4,095,900	3,203,500
202,500	176,600	237,700	273,200	298,900	309,000	360,000	398,700	362,500
33,400	42,221	76,816	106,300	99,109	107,300	127,000	132,900	106,200
834,596	744,500	794,600	958,600	957,100	477,701	792,700	801,700	554,600
162,780	253,549	464,600	600,800	683,700	720,301	807,170	908,000	819,000
650,800	655,300	617,800	614,100	620,700	589,700	607,500	512,700	353,200

such as cigarettes, cigars, rum, and gasoline. Thus, taxes collected on cigarettes rose about 89 percent between 1993 and 1994, staying at that new level through the rest of the decade; taxes collected on cigars jumped about 236 percent between 1993 and 1994 and hovered around the new level thereafter; and taxes for rum and gasoline grew approximately 100 percent and 69 percent, respectively (calculated from CEPAL 2000, A.19). Nonetheless, price increases also affected services such as the provision of electricity and water in residential areas, public transportation, and postage.

These various efforts by the government to rein in its budget deficit produced results. After rising from representing 9.4 percent of the GDP in 1990 to 30.4 percent of the GDP in 1993, this figure fell to 7 percent in 1994 and to 1.9 percent in 1996, before rising just slightly again toward the end of the decade.[24] This is not to suggest that the deficit was no longer an issue for government policy makers in Cuba but rather that the runaway deficit of the first half of the 1990s had been brought under control by multiple economic policies.

At the same time that subsidies were cut and the state's administrative structure was streamlined, both standard components of orthodox economic reform, another, rather unorthodox, approach to economic restructuring was promoted: the creation of linkages between the tourist sector and Cuban industry and agriculture. With the first major expansion in the tourist sector,

24. Calculated from CEPAL 2000, tables A.11, A.13; CEPAL/INIE/PNUD 2004, table I-1.

This favorable trend experienced a reversal in the 1998–2002 period (ONE 2003, table IV.4). Yet even in the worst of these years, the deficit was still approximately 50 percent less than it had been in 1990.

Table 5.4.2   Production levels by sector for Cuba's principal agricultural goods, 1990–2005 (tons)

	1990	1991	1992	1993	1994	1995	1996
State sector							
Root crops	483,338	503,746	566,863	420,296	182,187	228,360	269,141
Plantains/bananas	223,758	266,833	412,524	332,535	178,125	183,018	255,273
Vegetables	204,310	262,019	283,135	194,323	108,514	133,527	144,222
Corn	39,682	34,632	35,696	26,756	17,613	23,899	29,118
Beans	9,803	9,236	7,883	7,422	3,912	3,834	3,829
Citrus	922,273	736,102	688,837	560,815	329,154	346,101	401,481
Other fruits	109,918	124,792	65,623	37,121	24,185	22,880	23,486
Milk	820,300	629,900	365,100	316,700	167,600	117,700	123,200
	1990	1991	1992	1993	1994	1995	1996
Nonstate sector							
Root crops	218,952	186,700	187,079	148,431	302,350	395,835	473,150
Plantains/bananas	100,446	90,266	102,101	67,483	182,554	216,971	284,153
Vegetables	279,903	228,776	230,544	198,560	213,650	268,754	349,355
Corn	25,363	20,717	22,773	22,693	56,010	57,091	75,207
Beans	2,221	2,611	1,844	1,397	6,859	7,640	10,220
Citrus	93,600	89,899	98,143	83,651	175,837	217,438	260,720
Other fruits	109,050	132,840	61,799	31,224	64,925	89,410	79,068
Milk	214,100	190,400	257,200	268,900	468,000	520,800	516,800

Source: ONE 2007, tables 9.9–9.11, 9.17.
[a]Includes production of nonspecialized farms and home gardens.

and foreign investments more generally, some Cuban economists expressed concern about the increasingly dual nature of the economy (see Carranza Valdés, Gutiérrez Urdaneta, and Monreal González 1995). They saw the "new" sectors in which foreign investment was concentrating as monopolizing the dynamism within the economy and leaving behind the more traditional sector that was fundamentally oriented toward domestic consumption. Yet by the mid-1990s FINATUR—the Cuban financial institution that was established in 1992 to provide funding for tourism-related industries—had begun to invest notable quantities of resources in local industries to stimulate national production of the goods used in the tourist sector. From supplying an estimated 18 percent of the inputs required by this sector in 1990, local industry and agriculture were supplying 61 percent of them by 2000 (Marquetti Nodarse and García Capote 2002, 79).[25] This was important for two reasons: (1) it lowered the coefficient of imported inputs needed by this sector, thereby permitting Cuba to retain a larger percentage than would other-

25. CEPAL (2000, 538) states that national production supplied 10 percent of the input needs of tourism in 1990 versus 45–60 percent of them in 1998.

1997	1998	1999	2000	2001	2002	2003	2004	2005
245,681	191,950	252,030	252,291	268,150	229,418	261,469	292,767	235,487
193,965	241,334	205,055	200,192	209,733	145,584	212,320	226,701	167,384
163,450	347,098	377,765	498,137	587,075	708,919	693,058	742,567	635,783
34,411	37,594	35,547	35,039	38,780	36,606	39,961	33,999	28,201
4,628	6,564	6,053	9,125	7,398	9,626	10,580	13,030	11,091
453,293	376,279	358,778	520,361	472,098	108,031	373,729	464,460	287,629
26,228	37,176	71,091	58,377	65,401	67,326	93,374	93,867	83,875
132,200	126,600	106,900	102,300	107,800	87,800	90,500	111,500	69,200
1997[a]	1998[a]	1999[a]	2000[a]	2001[a]	2002[a]	2003[a]	2004[a]	2005[a]
581,601	625,620	807,217	978,550	1,112,409	1,207,611	1,582,131	1,653,633	1,566,313
335,335	324,766	398,145	644,708	758,267	584,316	900,280	988,899	606,116
437,571	499,711	1,064,741	1,874,536	2,089,400	2,635,791	3,238,142	3,353,333	2,567,717
168,089	139,006	202,153	238,161	260,120	272,394	320,039	364,701	334,299
28,772	35,657	70,763	97,175	91,711	97,674	116,420	119,870	95,109
381,303	368,221	435,822	438,239	485,002	369,670	418,971	337,240	266,971
136,552	216,373	393,509	542,423	618,299	652,975	713,796	814,133	735,125
518,600	528,700	510,900	511,800	512,900	501,900	517,000	401,200	284,000

wise be possible of the foreign exchange earnings generated within it; and (2) it helped to attenuate the tendency of the tourist sector to become an enclave, completely separate from the rest of the economy (see Enríquez 2007). Effectively, the promotion of industrial and agricultural production geared toward this sector—and toward supplying manufactured and agricultural goods to the government-owned stores that sold to the local population in foreign currency—represented the employment of an import substitution strategy in this crucial sector of the economy.[26]

This was not the only sector in which an import substitution strategy was adopted in the 1990s. The dramatic drop-off in imported inputs, including

26. Monreal (2002b), who was among those mentioned earlier as being concerned about tourism and related investments fueling the emergence of a dualistic economy, writing seven years later recognized the import substitution industrialization that was now associated with the tourist sector. However, in this later piece, he argued that this represented only a variation on Cuba's post-1970 development strategy. And, he asserted that for Cuba to truly embark on a new development trajectory it would have to replace its ISI strategy with one of export substitution industrialization. This would require a shift to exporting goods and services that involved manufacturing (instead of the sale of raw or semimanufactured goods) and that employed its labor force in ways that took advantage of its high skill level (rather than low labor costs).

agro-chemicals, for Cuba's agricultural sector provoked immediate shortages of these goods in the early 1990s. The shortages opened the way for the country's nascent organic agriculture movement to gain influence and support within policy-making circles, leading to the promotion of organic production (Funes Monzote 2002). Before this, as part of the predominant model of production, a reliance on agro-chemicals prevailed. With their reduced importation after 1990, however, came a whole new emphasis on using biological pest control and organic fertilizers (Rosset 1997; Wright 2005).[27] These biological inputs were produced locally in research centers scattered throughout the countryside to be accessible to farmers in all sectors of agriculture.

On an even grander scale, import substitution also took the form of efforts to move toward greater food self-sufficiency at the national level. The dissolution of the COMECON, which had previously ensured that Cuba's food needs were met, shifted into high gear the implementation of a food production plan that had been developed earlier but which had not yet been completely put into practice to meet the drastic shortfall in food imports that began in 1990 (see Enríquez 1994). Cuba's drive for food security in the 1990s thus was another distinguishing characteristic of the country's economic restructuring.

The Overall Impact of Cuba's Economic Reform

So what has the effect of this unorthodox reform been for Cuba's economy? I have already described how some of the measures that composed it achieved their desired effect (e.g., the fiscal deficit was cut significantly), while other measures did not yield the desired results (e.g., domestic food production had expanded in some crops, but contracted in others). Nonetheless, several other important economic indicators have not been discussed that can provide insight into the effect of the reform.

Cuba's GDP fell precipitously by 35 percent from 1990 to 1993. This extremely negative trend began its gradual reversal in 1994, with a positive 0.6 percent growth in GDP that year (CEPAL 2000, table A-1). The economic situation continued to improve during the remainder of the decade, with yearly variations in growth ranging from 7.8 percent in 1996 to 1.2 percent in

27. In fact, Pérez Villanueva (2004, n. 5) states that approximately 60 percent of vegetable production in 1998 took place on organic farms.

1998 (CEPAL 2000; Pérez Villanueva 2004a, 50). Although the growth rate was inconsistent, it was consistently positive (and both of these trends continued through the mid-2000s[28]), which demonstrated that the economy had emerged from the dramatic recession of the early 1990s.

Import and other data suggest that, despite the mixed results in food crop production, the import substitution strategy implicit in focusing agricultural resources on this type of production was having an effect. Between 1990 and 1994, food product imports dropped by 44 percent (calculated from CEPAL 2000, A.37), which could have reflected the disarray characterizing Cuba's trade relations following the demise of the COMECON and the country's inability to purchase imported goods because of its economic crisis. Food imports rose again after the 1994 low point, stabilizing in the latter half of the 1990s at roughly 86 percent of what they had been in 1990 (calculated from CEPAL 2000, A.37; CEPAL/INIE/PNUD 2004, A.42). Yet combining import and other kinds of data, Cuban researchers (e.g., García Álvarez 2004) have concluded that Cuba has, indeed, begun to substitute food imports with local production. For example, local production of beans and other legumes grew during the 1990s in relation to imports of them. Imported sources of calories decreased in relation to locally produced sources. Cuban agriculture has continued to fall short in the local production/import balance with regard to animal sources of protein. But progress has, at least, been made on some fronts.

Controlling inflation was another area in which the economic reform appeared to have made headway. Analysts of the Cuban economy use a few distinct indicators to describe inflation. Pérez Villanueva (2004, 56) and Togares González (2000, 7) suggest the pattern of inflation by illustrating shifts in the gap between the value of nominal versus real wages (the latter takes into account the consumer price index). In contrast, Domínguez (2004, 19) estimates inflation based on a GDP price deflator. Despite these differences, both avenues of calculation lead to the conclusion that inflation was severe between 1990 and 1995, with a GDP price deflator in the latter year of 157.2 (1985 = 100 [1985 served as the base year]; Domínguez 2004, 19), and that prices stabilized thereafter. In fact, Domínguez (2004, 19) notes that the GDP price deflator "remained basically unchanged" throughout the rest of the decade.[29]

28. See Economics Press Service (2007), Pérez Villanueva (2004), and CEPAL/INIE/PNUD (2004).

29. Both of these methods of calculation fail to consider shifts in black-market prices, thereby underestimating inflation. But the trend they point to is arguably correct, as the greater

The picture was less optimistic regarding the country's foreign trade balance. The trade deficit dropped by approximately 90 percent between 1990 and 1992, as both exports and imports underwent a drastic contraction (CEPAL 2000, table A.30). Once trade started to pick up again, however, exports failed to keep pace with imports. The average deficit in the second half of the 1990s was less than a third of what it had been in 1990.[30] But what this really underlined was that exports had not grown to the extent they needed to in order to ensure a full economic recovery (see Table 5.5).

The principal export product in Cuba's repertoire since the late 1700s had been sugarcane, and this continued to be so through the end of the 1990s and into the twenty-first century (see Table 5.5). Tourism had supplanted sugarcane in generating foreign exchange earnings by 1994. Yet sugar was still central to the country's economy.[31] Therefore, its dramatic decline in earnings through the 1990s and the first half of the 2000s had implications for the entire economy. Its decline was a result of the growing production cost/price squeeze experienced by producers all over the world (cf. Peña Castellanos 2002, especially figure 5.4). This squeeze had a particularly strong effect on Cuba because, as stated earlier, before 1990, it had received preferential pricing through its participation in the COMECON. Moreover, by 1998, oversupply at the world level set in motion a downward trend in the price for sugar that, it was presumed, would go on at least for the medium term. In addition, "technological obsolescence and low agricultural and industrial yields" contributed to Cuba's sugar woes in the 1990s (see Peña Castellanos 2002, 96). Both of these latter problems were closely associated with the general shortage of resources prevailing in the country in the 1990s but also to the lack of priority given to this specific sector.

By September 1993, the massive drop in sugar production had led to the dismantling of the state farm sector. External financing facilitated investments that resulted in an improvement in earnings in 1996 and 1997, but a consistent upward trajectory had yet to be obtained. Finally, in mid-2002, government policy makers decided that even more radical measures would have to be taken. As a consequence, close to half of the country's sugar

---

availability of goods that emerged in the mid-1990s—brought about by the opening of the Mercados Agropecuarios, among other things—succeeded in bringing down black-market prices.

30. CEPAL (1990, table A30); and CEPAL/INIE/PNUD (2004, table I-17).

31. This was especially so with regard to employment generation. In 2002, Peña Castellanos (2002, 97) noted, "Today, as in the past, the Cuban sugar industry directly employs approximately 460,000 people, and indirectly some 1.5 million, accounting for close to 50 percent of the country's total available labor force."

refineries were shut down to concentrate production in the most productive areas/mills. The mills that were slated for closure were supposed to become food-processing plants, and the associated farmland was to be used for growing food crops and timber and raising cattle. Approximately 100,000 workers were to be reassigned to the remaining mills, to the new food-processing centers, or to other farm jobs (González 2002). The ramifications of this decision were social, cultural, and economic.

There were other social effects of Cuba's economic crisis and reform, however. These included, as suggested in Chapter 1, an increase in inequality within the population. This increase can be seen in a variety of ways. One way is through looking at the Gini indices that describe inequality. In 1978, after nineteen years of implementing policies designed to reduce inequality in Cuba, the country's Gini index was .25 (Brundenius 1984, 116).[32] By 2002, this figure was estimated to be .38.[33]

Another way of assessing inequality is by looking at the gap in income between those who earn the most and those who earn the least. By the mid-1980s, most income in Cuba was received in the form of salaries, through a specified system of remuneration. Within that system, the highest earners' salary was approximately 4.5 times that of the lowest earners (Espina Prieto 2004, 219). Nonetheless, qualitative studies in 2001 and 2002 suggested that this gap had grown exponentially, with the higher-income earners perhaps receiving as much as twenty-eight times that of the lowest-income earners (Espina Prieto 2004, 222).[34]

Inequality was also growing between regions. Regional inequalities had their roots in the differences that emerged between them in terms of their economic development, which will be described further in Chapter 6. Yet the 1990s witnessed a renewed process of regional differentiation in economic inequality. Quintana Mendoza (1996, 1998), who analyzed this process as it was unfolding, noted that little change occurred in terms of regional differentiation in salaries among those employed by the state. Instead, increasing inequality between regions could be seen in shifts in savings and in other forms of liquidities. After the imposition of the various austerity measures, especially the price increases for certain goods and utilities discussed

32. Espina Prieto (2004, 219) states that the Gini index was .24 in 1986.
33. Ferriol (2002), as cited in Espina Prieto (2004, 221). See also, Ferriol Muruaga (2000, 41), who states that the Gini index for Cuba's urban areas in 1996–98 was .38.
34. On the basis of interviews with Cuban visitors abroad and Cuban émigrés, Mesa-Lago (2003, 79) came up with much higher differences: 829 to 1 for 1995; and 12,500 to 1 for 2001.

Table 5.5   Value of Cuba's exported goods, 1990–2005 (thousands of pesos)

Product	1990	1991	1992	1993	1994	1995	1996
Food products and live animals[a]	4,640,715	2,509,128	1,385,208	867,612	914,523	899,581	1,168,389
Sugar products	4,337,500	2,287,500	1,240,200	758,100	759,500	714,300	976,300
Beverages and tobacco products	127,650	119,643	100,431	79,398	78,708	112,098	122,151
Nonfood raw materials	410,969	254,247	242,544	164,449	208,845	351,784	437,605
Chemical products	92,239	44,235	17,030	7,078	77,564	53,131	59,899
Others	143,376	52,259	34,211	38,126	51,116	75,040	77,482
Total	5,414,949	2,979,512	1,779,424	1,156,663	1,330,756	1,491,634	1,865,526

SOURCES: ONE 1998, tables VI.8, VI.10; ONE 2003, tables VI.8, VI.10; ONE 2006, tables VII.7, VII.9.
[a]Includes sugar products.

earlier, savings in the eastern part of the country—the Oriente—dropped dramatically compared with those in the western part of the country. That is, while those from the western part of the country were able to supplement their salaries with other income sources, those from the Oriente were increasingly forced to rely on savings to survive. Having access to earnings in U.S. dollars was the key factor in explaining this difference.

What became apparent during the 1990s was that inequality between regions also existed with regard to access to earnings in U.S. dollars, with the Oriente being disadvantaged in this sense. This was probably due to at least a few distinct factors. Among these was that the western part of the country, especially the City of Havana, tended to contribute more heavily to emigration from Cuba after 1959 than did the Oriente. This had a direct bearing on receipt of remittances (Eckstein 2004, 342).

A related factor was the racial breakdown of those who emigrated versus those who stayed. As Eckstein (2004, 342) notes, among Cuban émigrés who ended up in the United States, most (84 percent) identified themselves as white. Therefore, the percentage of white Cubans on the island who received remittances was substantially higher than Afro-Cubans.[35] Given the larger presence of Afro-Cubans in the Oriente, the effect of the differential receipt of remittances would play out in regional income differences.

However, Martín et al. (1999, as cited in Espina Prieto 2004) also highlight the importance of provinces' linkages with the newly "emerging econ-

35. Mesa-Lago (2003) reaches the same conclusion.

1997	1998	1999	2000	2001	2002	2003	2004	2005
1,043,850	776,262	659,992	656,703	711,097	594,560	439,689	459,746	—
853,300	599,300	462,500	452,600	550,300	448,100	288,600	271,500	—
173,796	202,562	218,471	179,719	232,023	162,373	238,231	243,597	251,403
427,143	358,781	436,808	623,089	485,416	463,237	655,450	1,110,096	1,039,275
55,662	45,566	39,662	41,227	56,168	65,033	60,813	51,667	103,928
118,676	129,026	140,850	174,521	137,187	136,455	277,448	1,322,896	—
1,819,127	1,512,197	1,495,783	1,675,259	1,621,891	1,421,658	1,671,631	2,188,002	1,998,670

omy" in determining this territorial restratification. The newly emerging economy was based in the tourist and joint venture sectors. The Oriente had only limited participation in these sectors. This dynamic was reflected in, and probably affected by, as well, the fact that fewer Afro-Cubans obtained employment in the tourist sector than whites (cf. Eckstein 2004; Mesa-Lago 2003).

At the same time, the Oriente's previously existing economic structure, particularly its industry, was extremely hard hit by the dissolution of the COMECON and the ensuing economic crisis. Its production had been partially directed toward export to the COMECON and partially directed at the local market, both of which ceased to be a priority in the 1990s. Consequently, such industries were shut down. One gigantic multiproduct factory alone, the "Celia Sánchez," which had employed approximately 10,000 people in its heyday, closed its doors in the early 1990s because of the shift in industrial policy that was part of the economic reform (Interview, government planner—2 July 2001). The Province of Santiago was left to somehow absorb this huge pool of newly unemployed people. These kinds of plant closings, far removed from the more dynamic sectors of the economy, were bound to have an effect on the local population and its income-generation potential.

Cuba's economic crisis and the reforms it gave rise to also led to growing poverty. Although there is some debate about the extent of poverty that existed before 1990, the salary system that was in place and the extensive social service coverage had greatly reduced its significance, if not eliminated it entirely, as a social problem. Yet the population at risk of poverty grew by 138 percent between 1988 and 1996, increasing from 6.3 percent to 15 percent (Ferriol Muruaga 2003, 14).[36] Ferriol Muruaga (2003) posits that this latter figure actually represented an improvement—resulting from the economic recovery that was underway by this time—from what it must have been when the country's economy hit bottom in 1993. Parallel to regional differentiation in earnings, poverty had also come to affect a wider pool of people in some regions than others. The urban population considered at risk in the Oriente in 1996 reached 22 percent (Ferriol Muruaga 1998, 12).

Despite Ferriol Muruaga's (2003) suggestion that poverty rate trends had improved after 1993, more recent data highlight that the population at risk continues to be an issue. In 1999, using a somewhat different system of calculation, thereby making comparison complicated, it was estimated that about 20 percent of Cuba's urban population was at risk of poverty (Ferriol Muruaga 2003, 14). It is crucial to underscore that poverty levels in Cuba are among the lowest in Latin America (see CEPAL/INIE/PNUD 2004, 66, 68; CEPAL 2007). As important, inequality in Cuba remained the lowest in Latin America and the Caribbean. Nonetheless, it is still essential to acknowledge that increased poverty and inequality were among the effects of Cuba's economic crisis and reform.

Conclusion

Cuba's economic reform evolved over a period of years, beginning in 1990/91 with the effort to diversify and expand the country's foreign economic relations. Following that more limited approach to reform came a deepening of it in key aspects of the domestically oriented economy and overall economic structure. In some of these aspects, Cuba's reform resembled those

36. The term *at risk*, used when describing poverty levels in Cuba, is meant to acknowledge that even though income data would place the indicated population in this category, they had a notably distinct social safety net (through the food rationing system; free, high-quality public education and health care; etc.) than those who fell in this category elsewhere in the Global South (see Ferriol Muruaga 2003, 8).

However, Mesa-Lago (2003) disagrees with the methodology for calculating poverty that was employed by Ferriol Muruaga. He argues, instead, for a modified version of the methodology employed by another Cuban economist, Viviana Togores. His modification does not consider the counterbalancing influence of state-provided social services that Togares stipulates must be included in this equation. Using his modification to this methodology, he concludes that poverty actually affected between 61 percent and 67 percent of the population (Mesa-Lago 2003, 108).

imposed elsewhere. Thus, there was a clear opening to the international economy and the partial liberation of market forces within the domestic economy. Moreover, as in more traditional economic reforms, subsidies were cut in many parts of the economy and a streamlining of state enterprises and the state administrative structure was carried out.

These modifications in economic policy had social consequences, as seen in the rise in poverty and the increase in inequality. The social consequences were consistent with what scholars have found elsewhere when structural adjustment has been implemented in the contemporary era (see Bulmer-Thomas 1996; Korzeniewicz and Smith 2000), as well as with what Polanyi (1944) described for the first era of liberalism. They also reflected what Szelenyi and Kostello (1996) had suggested would occur with the expansion of market relations within the formerly socialist economies.

Yet the fact that Cuba continued to have lower levels of poverty and inequality than elsewhere in Latin America suggested that some elements of a "redistributive" framework remained hegemonic (à la Szelenyi and Kostello). That is, what took place in Cuba after 1990 was a reconfiguration of socialism rather than a retreat from it. Cuba's unique transition toward the market was also expressed in the state's enduring central role in the economy. Even with the transformation of many state farms into UBPCs, the "privatization" process, as it has been called by some (see Deere 1994), was only partial and the state was still a major actor in planning, production, and distribution.

One sector of the economy in which competing approaches were especially palpable was agriculture. Post-1990 agricultural policies point in a distinct direction from both what socialist agriculture had been before that time in Cuba and from what proponents of economic liberalization typically envision this sector to look like. It is to an examination of the reshaping of socialism within this sector, and its effect on the country's small farmers, that I turn in the following chapter.

6

The Reconfiguration's Varying Impact on Cuba's Small Farmers

The reconfiguration of Cuban socialism set in motion in 1990 had a myriad of facets. In the agricultural sector, the downsizing of production, the liberalization of marketing, the turning over of parcels of land to those willing to work them, and so forth, all formed part of it. These diverse policies were geared to incentivize production increases, incorporate underemployed labor, and eliminate inefficiencies in production, especially in light of the massive drop-off in imported inputs that had made large-scale production much more difficult. Success on these fronts would help to relieve the country's food crisis for consumers in urban areas as well as for those engaged in farming, at the same time as contribute to the larger effort to improve Cuba's macroeconomic situation.

The new agricultural policies were all national-level initiatives. Thus, it was to be expected that, to some degree, their results would be similar everywhere. Yet because of differences in geographies, historical patterns of production, and levels of development, it could also be supposed that some variations in the effect of the political economic changes of the 1990s might emerge between regions. Given my interest in gaining a comprehensive sense of the ways in which the livelihoods and lives of small farmers had been altered by this transformation, it would be essential to examine their circumstances in several distinct parts of the country.

With this objective in mind, in 1998, I interviewed small farmers, whose specialty was food crop production, in four largely rural municipalities in

Cuba. These municipalities were divided evenly between two provinces that contrasted notably in terms of geography, level of development, and the socioeconomic status of their populations. The first of these was the Province of Havana. This province adjoins the capital. Its fertile soils cover the plains that surround the capital on Cuba's north coast and extend to its southern coast. Its proximity to the City of Havana (which is a separate jurisdiction) and topography undoubtedly contributed to its high level of infrastructural development. This, in turn, surely facilitated the adoption of more advanced levels of technology in both agricultural and industrial production there. The two municipalities included in my study in this province—Güira de Melena and San Antonio de Los Baños (see Map 2)—were selected because they both contained all of the organizational forms of production I wanted to examine.[1]

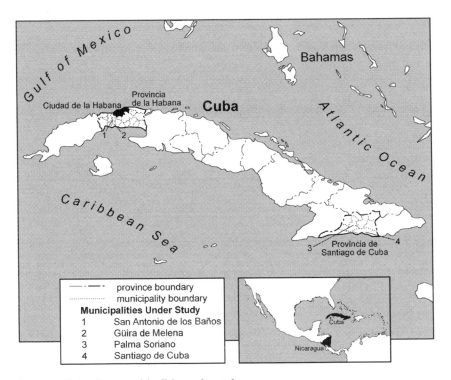

Map 2   Cuba: Four municipalities under study

1. I wanted to interview individual *parceleros*, as well as members of CCSs (to be defined later), CPAs, and UBPCs. I decided to include people from all of these sectors within my defini-

The second province I conducted interviews in—Santiago de Cuba—is situated in the Oriente. This province is mountainous, making it ideal for the production of certain kinds of crops, such as coffee, but less ideal for many basic food crops.[2] Although the city of Santiago has a port, its significance bears no relation to that of Havana. The province, like the rest of the Oriente, has been characterized by a lower level of economic development and standard of living, and throughout the second half of the 1990s was affected by a severe drought. The two municipalities in which I conducted fieldwork in this province were Palma Soriano and Santiago de Cuba (see Map 2).[3]

I will begin by describing the situation of farmers from the four municipalities together. Later, I will discuss the principal differences I encountered in their experiences.[4] The comparison will explicate the general effect of

---

tion of "small farmers" in Cuba because the area of individual farmers is limited enough in size to qualify them as such; and, if the area of a cooperative were to be divided among its membership, they too would qualify as small farmers.

Within Cuba, there is disagreement concerning exactly how broadly the term *small farmer* (*campesino*) should be defined (Martín [1998] describes the differing scope of these various definitions). Some would limit the use of this term to only members of CCSs and CPAs (that is, individual property owners and cooperative members). Others argue for a more inclusive definition, which incorporates *parceleros* and UBPC members as well, while acknowledging that those from this last group are a transitional sector of sorts (moving from being agricultural workers to small farmers). I have opted to use this broader definition.

As to the precise size specifications, in areas that are fertile, strong in terms of infrastructure and level of technology employed, and relatively close to markets (e.g., the Province of Havana), small farmers have less than fourteen hectares. This corresponds to my specification of the term *small farmer* for León, Nicaragua, which roughly shared these characteristics.

In addition to wanting to include farmers from these various organizational forms, a couple of other factors were influential in my selection of these municipalities: I had some previously established contacts in one of them, and logistical issues also came into play because of the tremendous difficulty of housing, food, and transportation in the Cuban countryside in the 1990s, including as they affected this researcher.

2. For purposes of my study, small farmers in the Province of Santiago de Cuba were defined as those who had less than thirty-five hectares. This was how I specified the term *small farmer* more generally in areas that were less fertile, more limited in infrastructure and level of technology, and further from markets. This corresponds to my specification of the term *small farmer* for Matagalpa, Nicaragua, which roughly shared these characteristics.

3. I selected these two municipalities for similar reasons to those mentioned in n. 1 for the Province of Havana. That entrée to these particular municipalities was made possible for me was also a crucial factor.

4. As the reader may have noted, the structure of this chapter is distinct from its comparative case, Chapter 4 on Nicaragua. Because the type of economic reform pursued in Nicaragua was more traditional and, therefore, more familiar, my description of it could be less lengthy in Chapter 3. This allowed me to include a discussion of the reform's effect on agriculture more generally in that chapter. Consequently, Chapter 4 could be entirely devoted to examining its differentiated impact.

Cuba's economic crisis and the multiple reforms that were employed to try to rein it in, as well as the ways in which they interacted with regional distinctions to produce varying outcomes.

Small Farmers' Access to Agricultural Resources

As described in Chapter 4, access to productive resources defines the well-being of farmers everywhere. Given Cuba's economic crisis and the resulting shifts in economic policy, it is hardly surprising that productive resources were a serious concern for its farmers after 1990. However, the specific resources that were in limited supply in Cuba differed from those that are typically a problem for small farmers elsewhere, including in Nicaragua.

Where thoroughgoing land reform has not been undertaken, land is usually the resource most lacking within this sector of the population (see Thiesenhusen 1995; de Janvry 1981; Murdoch 1980). But, because of the variety of forms through which land was available in Cuba, it was largely taken for granted. With the agrarian reforms of 1959 and 1963, and the possibility of joining an Agricultural Production Cooperative (CPA) once the cooperative movement took off in the late 1970s, the avenues to gain access to land were multiple even before 1990. Since 1990, through the establishment of the Basic Units of Cooperative Production (UBPCs) and the parcelization of some state farmland, even more avenues were opened up. In fact, only one of the fifty-nine producers included in my survey in Cuba, a *parcelero*, mentioned insecure land tenancy as a preoccupation.

Likewise, agricultural credit has traditionally been hard to come by for small farmers all over the world. With the widespread implementation of structural adjustment programs, access to this always scarce resource has become only more constrained[5]—as was the case in post-1990 Nicaragua. Yet in Cuba, credit was another taken-for-granted item for most farmers I interviewed (see Table 6.1). Credit was readily available from the Banco Nacional de Cuba. Moreover, interest rates on production loans were extraordinarily low at 4 percent. Nonetheless, most of those interviewed did not use credit,

---

In contrast, given Cuba's unorthodox approach to economic reform, it was necessary to describe its overall nature at greater length in Chapter 5. Thus, I have dedicated more space to a discussion of its general impact on agriculture, aside from the production levels that were described in Chapter 5, herein, before presenting a somewhat briefer depiction of regional differences.

5. Jonakin and Enríquez (1999), Carter and Barham (1996), and Serra and Castro (1994), among others, speak to this issue.

Table 6.1   A comparison of four municipalities: Changing access to agricultural resources and its effects on small farmers (percentages)

| | Province of Havana | | Province of Santiago | |
	Güira (N = 15)	San Antonio (N = 16)	Palma Soriano (N = 15)	Santiago (N = 13)
Land				
Owned (inherited)	26.7	37.5	20.0	30.8
Agrarian reform	73.3	62.5	80.0	69.3
Before 1990	27.3	60.0	16.7	22.2
After 1990	72.7	40.0	83.3	77.7
Credit				
Used it	73.3	31.3	40.0	23.1
Did not use it	26.7	68.8	60.0	76.9
No need	75.0	100.0	88.9	90.0
Technical assistance				
Received regularly	100.0	100.0	93.3	84.6
Didn't receive regularly	—	—	6.7	15.4
Increased self-sufficiency	46.7	18.8	53.3	30.8
Of those who received land post-1990	(N = 8)	(N = 4)	(N = 10)	(N = 7)
Came from agricultural sector	50.0	—	30.0	—
Came from urban sector	50.0	100.0	20.0	71.4
No answer	—	—	50.0	28.6
Improvements on the farm				
Were possible	66.7	81.3	60.0	76.9
Purchased animals	53.3	37.5	26.7	38.5
Forced sales of animals/equipment	13.3	—	—	—

SOURCE: Author's survey data.

either for investments or for ongoing production costs (only 42.4 percent used one or both types). The vast majority of those who did not use it (91.2 percent) stated that they simply did not need it.[6] They were able to set aside enough in savings after each harvest to cover any production-related expenditures in the next agricultural cycle.

Another resource that can be highly beneficial in farming is technical assistance. In contrast to the limited access to technical assistance that most small farmers around the world experienced, it appeared to be readily available to Cuba's small farmers. Approximately 95 percent of those included in my survey regularly received technical assistance from either state extension workers or those associated with their cooperative (see Table 6.1). Thus, as

6. Although it should be mentioned that some cooperative officials indirectly expressed disapproval of farmers who continued to rely on credit, I believed the sincerity of those who said that they had no need for it and therefore did not use it.

these farmers began to experiment with growing a variety of new crops and raising farm animals, which will be described later, they would appear to have received crucial technical support for making such changes successful.

However, the availability of a set of resources was somewhat constrained for Cuba's small farmers—production inputs. Small farmers elsewhere may be restricted in their access to the variety of inputs that can enhance their production, particularly because of their lack of purchasing power. Yet until 1990, most small farmers in Cuba benefited from their relative abundance. The range of inputs they had employed was comparatively large, reflecting the different levels of development of the productive forces in the country. By and large, small farmers there were not held back in their input use to anywhere near the same degree as in other countries.

As of 1990, though, Cuba's access to agricultural inputs dropped off dramatically. As described at length elsewhere,[7] imports of pesticides, fertilizers, diesel fuel (used in irrigation systems and by farm vehicles), and farm equipment plunged after 1990, which inevitably affected farm production. But, with the country's economy beginning to rebound by the late 1990s, limited access to inputs was not (or was no longer) an all-pervasive problem for small farmers. Although many of those who participated in my survey mentioned the restricted availability of inputs, only 40.7 percent said that it was affecting their production levels.[8] The remainder were able, through various means (including *guapeando*[9]), to resolve this problem. For many, especially from the CPA sector, input shortages brought about a more efficient use of resources. The use of oxen and organic inputs (both of which rely on local resources) increased, which also helped to alleviate the effect of shortages of imported inputs.

Post-1990 Changes in Production and Marketing Patterns

My survey did, however, reveal several important changes in production patterns that appeared to be largely related to the existence of the Special Period. The most notable modification was that Cuba's small farmers,

7. See Deere (1991), *Granma* (1991), and Enríquez (1994).
8. Several of those who fell in this latter category specified how long input shortages had been an issue for them. Among those who made such a statement, 75 percent indicated that this had only been so since 1990.
9. In Cuban vernacular, *guapear* means "to be tough." But, in this context, it means "toughing it out."

regardless of their particular production emphasis, had begun to set aside part of their cultivable land for crops that would ensure that their own, their family's, and their employees' consumption needs were met. Given a long history of extreme crop specialization,[10] the adoption of more integrated production patterns was significant. Even those farmers whose specialty had been a food crop or two were producing a much greater variety of crops at the time of our interview. And, those whose production emphasis had been on export crops had begun to grow food crops. More than 37 percent of the survey informants spoke explicitly about undertaking this change since 1990 (see Table 6.1).

The introduction of livestock was another aspect of the new stress placed on integrated production. Most Credit and Service Cooperative (CCS) members—those farmers who continued to own and farm their land individually but joined together to receive certain resources and, at times, to sell their produce—had always kept some small farm animals for their consumption. However, by 1998, 89.8 percent of those interviewed raised some livestock on the land they worked. While not all of them kept a wide array of animals, most at least had pigs and chickens, if not also cows and sheep.

These changes reflected both the government's new emphasis on ensuring some degree of self-sufficiency in the countryside, as well as the farmers' own concern not to be dependent on the official food distribution system or even on the Mercados Agropecuarios. As one CCS member mentioned in 1994, having such a self-sufficiency plot was essential for, among other things, ensuring the farm a stable workforce.[11] It was the principal means of attracting workers at the time because a salary alone was not enough to secure a reasonable food supply. Likewise, when I asked those farmers who had undergone the transformation from being urban workers before 1990 to being farmers in the 1990s (25 percent of my sample) why they had done so, having guaranteed access to food was mentioned by a number of them.[12]

One further change in production patterns brought about by some of those interviewed may also have been related to the circumstances of the

10. Pérez Marín and Muñoz Baños (1992) and Rosset (1997).
CCS members had always produced a greater variety of crops than farmers from other sectors. But, after 1990, they began to grow food crops for their employees and extended family relations, as well, and there was greater diversity in their production.
11. Interview, CCS member, Güira de Melena, May 21, 1994.
12. IPF/FPNU (1996a) found the same to be true in their research on UBPCs. And, Deere (2000) found that self-provisioning was also key in the growing interest of workers (including state farmworkers) to join CPAs.

Special Period. Ten percent of those included in the survey mentioned no longer growing a certain crop or raising certain farm animals because they were an attractive target for thieves. When they were asked about the incidence of theft of crops and farm animals, a number of farmers replied that the problem was no worse than it had always been; others talked at length about how serious it had become in recent years. The crops/livestock most frequently dropped from the roster of those grown/raised on the farm for this reason were: plantain, yucca, cows, and oxen.

Finally, as mentioned in Chapter 5, in the mid-1990s, the government began to offer incentives for the production of selected crops. A number of the crops that these farmers produced were included in the list of incentivized products. Thus, on two cooperatives I visited—from which I drew five people for my survey—tobacco production was adopted in the second half of the 1990s. These cooperatives had heeded the government's call—and incentives—to try out its production.

Another area in which a major change had occurred for these farmers was in marketing their crops. Given that it became legal in 1994 for farmers to sell their excess production in the Mercados Agropecuarios, those I interviewed sold their produce in a variety of places. As decreed by law, most (93.2 percent) farmers in my survey sample still sold the bulk of their produce to Acopio. Once having complied with the sales quotas set by Acopio, however, most (83.1 percent) opted to sell the remainder of their produce in the Mercados. They sold anywhere between 15 and 100 percent of their produce there. In those cases in which it was this higher figure, their sales typically consisted of items that state agencies did not purchase. The norm was between 15 and 40 percent of their produce. And, some informants sold their excess production directly to people who came to their farms to make purchases.

But there were also farmers who, instead of selling their produce to Acopio, sold it to another state agency. Cítricos Ceiba and the Empresa Comercializadora were two such agencies. Both of these purchased better quality goods at higher prices than Acopio, and either exported their produce or sold it in the tourist sector of the economy. Moreover, their prices were often attractive enough so that farmers would sell their entire marketable crop to them.

Even though virtually none of these farmers sold all of their produce to Acopio, some of them were reluctant to acknowledge sales in the Mercado Agropecuario. The issue of how much of their produce they sold there was even more delicate. Clearly, a concern about the state overturning its decision to permit such sales, such as occurred in 1986 (see Deere and Meurs 1992; Rosenberg 1992), still remained among some farmers at the time of our interview. They may also have been strongly encouraged to sell all of their produce to the state and were loath to admit that they were doing otherwise. It would appear that more time would have to pass, with the Mercados remaining a legal option for them, for Cuba's small farmers to feel completely at liberty to sell their produce there.

Selling some of their produce at market prices, however, was only one means these farmers used to boost their incomes. Almost a third of them (28.8 percent) engaged in some other income-earning activity in addition to their farmwork. The other activities were agriculturally related: they included raising chickens, pigs, and ducks for sale; tending to fruit trees at their houses (among CPA or UBPC members); and providing services with their own oxen for pay. These activities involved farmers from all of the nonstate sectors of agriculture but were especially concentrated among the CPA and UBPC members.

Searching for extra income in hard economic times is certainly not unique to these informants. Chapter 4 described the prevalence of this phenomenon among Nicaraguan farmers in the 1990s. Perhaps the key distinguishing characteristic of Cuba's small farmers is that all of these supplementary activities were agriculturally related. Elsewhere, small farmers also engaged in production of artisan goods, sold their labor in and outside of agriculture, and so forth. In contrast, Cuban agriculture offered them means of generating supplemental income without their having to resort to wage labor or to economic activities outside of this sector.

New Structures and Forms of Agricultural Production

As described in Chapter 5, since 1990, a variety of modifications have been made to the model of agriculture that prevailed in Cuba before that time. These changes were in evidence to differing degrees throughout my interviews. They suggest trends that may hold into the future, while also revealing some interesting social dynamics in the present.

The downsizing of agricultural production was one of the more far-reaching changes. The most dramatic shape this took was the formation of the UBPC sector of agriculture. Through the inclusion of several members of the recently established UBPCs in my survey sample, I was able to identify some patterns among their membership, as well as within the sector. UBPC membership was largely made up of those who had worked on the state farms from which they were formed. However, my interviewees were almost evenly divided between those who had been regular workers on the *granjas* and those who had been Contingente members (to be discussed later).

More than half of the UBPC members I interviewed were migrants from another part of the country. Within this group, the obvious area of "pull" was the Province of Havana, while the obvious area of "push" was the Oriente. Thus, these new UBPC members embodied one of Cuba's most notable (if not *the* most notable) internal migratory patterns of the post-1990 period. While emigration from the Oriente was not a new phenomenon, it skyrocketed as of 1992.[13] The westward migration was, undoubtedly, related to the differential impact of the economic crisis and reform. This was manifested in the fact that the Oriente had the highest concentration of unemployment in the country.[14] Its relatively high unemployment contributed to the deepening of historic disparities in earnings that occurred between these two regions during the 1990s. As a consequence, people moved in search of employment and better economic conditions.[15]

Yet, the Province of Havana also "pulled" these migrants to it, with its need for agricultural laborers. The shortfall in inputs to fuel the province's (and the western and central regions' more generally) highly mechanized and technology-reliant agriculture, resulting from the imports crisis that emerged in 1990–91, created an increased need for people to work the land.[16] It was

13. The 1995 National Survey of Domestic Migration found that 47.5 percent of the population that participated in this phenomenon hailed from the Oriente, with the Province of Santiago de Cuba alone as the birthplace of 13.2 percent of them (IPF/FPNU 1996c, 27).

14. According to an IPF/FPNU (1996a, 28) study, 52 percent of Cuba's unemployed workforce was located there; the City of Havana was a distant second with 20 percent of the unemployed.

15. In fact, emigration from the Oriente to the capital, which was grappling with its own unemployment problem (as noted above), grew so much during the mid-1990s that the government decided that the flow had to be stanched. Thus, it imposed new regulations on migration to the capital as of 1996 in the form of Decree 217, "Regulaciones Migratorias Para la Ciudad de La Habana."

16. Labor shortages in agriculture first emerged in the post-1959 period as the agrarian reform gave land to the landless and the new emphasis on providing education for all opened up opportunities that took growing numbers of people out of the agricultural labor force. In fact, between 1953 and 1981, urban dwellers went from representing 57 percent of the population to 69 percent, and this trend continued through the 1980s and early 1990s (IPF/FPNU 1996a, 24). Mechanization was seen as a key solution for these labor shortages, and reliance on it increased significantly from 1959 to 1990. Yet efforts had also been made, through government-financed home construc-

calculated that the nationwide deficit for agricultural laborers was on the order of 210,000 people in the early 1990s (IPF/FPNU 1996a, 27).

The government's response to this labor-shortage crisis was multifaceted. It organized the large-scale mobilization of students and urban workers to participate in agricultural labor on a temporary basis.[17] At the same time, it increased its efforts to entice the urban unemployed, whether from the City of Havana or elsewhere, to commit to a two-year stint as a Contingente member working in agriculture. The offer of free housing (always in short supply in Cuba), comparatively high salaries, and other benefits were extended to them. The objective was to convince them to resettle permanently once they had seen that rural life had certain advantages. Then, this force of Contingente members became transformed into UBPC members with the dismantling of the state farms they had worked on. And, they, as well as the *parceleros* and new recruits to the CPA sector, came to replace the mobilized students and workers during the mid-1990s. This is not to suggest that the labor deficits had been resolved by this time.[18] Rather, it represented a recognition that solutions other than the temporary mobilization of less productive urban dwellers needed to be found, such as the longer-term relocation of people to work in agriculture, including through the encouragement of migration from the economically depressed Oriente to the rural parts of central/western Cuba. Within my overall survey sample, it was among UBPC members that migration from one part of the country to another was highest.

With regard to the UBPCs, I asked those from this sector whom I had included in the survey sample what they saw as the differences between their UBPCs and cooperatives. Most responded that there were none.[19] But, I

---

tion and the provision of many social services and other amenities, to make rural life more attractive than it had been.

17. See further Deere (1991) and Enríquez (1994).

18. In fact, IPF/FPNU (1996a, 53) found that 74 percent of the sugarcane UBPCs were still experiencing deficits in the mid-1990s.

19. Here, it is crucial to note that all of the three UBPCs that I drew members from to include in my survey were profitable enterprises. However, as discussed in Chapter 5, this was not the norm for UBPCs in 1998. Thus, it must be borne in mind that my informants were members of less than typical UBPCs.

In addition, a phenomenon widely commented on by those studying the UBPCs is that there is a relatively high turnover rate among their members (Torres Vila and Pérez Rojas 1996a; Bu Wong et al. 1996; IPF/PFNU 1996a). This calls into question the degree of commitment they feel toward the UBPCs and, hence, the extent to which they consider themselves to be the "owners" of their land (see Villegas Chádez 1996). This sense of ownership is crucial for the consolidation of cooperatives. See also Deere (2000) on this topic.

found several other responses to be quite telling. These described (with greater or lesser degrees of positiveness) the close relationship that continued to exist between the UBPCs and the remnants of the state farms that they had previously formed a part of. That is, although the UBPCs were supposed to have a considerable degree of autonomy from the state—because they were now "cooperatives"—their ties to it remained strong (see also Averhoff Casamayor 1996; and IPF/FPNU 1996a).

The downsizing of agriculture that the UBPCs embodied, however, also took the shape of subdividing production responsibilities on the UBPCs and CPAs, as described in Chapter 5. I found this phenomenon of "linking the person with the area" to be quite widespread during my fieldwork in 1998. Even though not all of the CPAs and UBPCs I visited had organized their membership in this way (either partially or totally), 81.8 percent had.[20] For the better part of those working on a CPA or UBPC where such subdivision had occurred, the economic benefits they received were notable.[21] In the eight cases in which the members mentioned the income they earned from this *vinculación* category of payment, the amount of additional earnings they received was equivalent to 7–80 percent of what their earnings from the "salary" category of payment were (the median was between 19 and 35 percent). Thus, where a need to boost production existed, linking the person to the land had the potential to facilitate it by providing cooperative members with a tangible incentive to exert themselves in a specific area of the cooperative where their efforts would be recognizable.

The effort to boost production also took the form of expanding the pool of people involved in farming through the creation of the *parcelero* sector, as mentioned in Chapter 5. Given that the individuals who make up this sector are, by definition, small farmers, and that their existence is emblematic of the post-1990 changes in Cuban agriculture, I included several *parceleros* in my survey sample.[22] Almost all of the *parceleros* I spoke with (nine were included in the survey sample) worked full time on their land. In the only case where this was not true, in addition to working his plot, the informant provided services for payment with the oxen that he owned. Before becoming a *parcelero*, this individual had worked on the state farm that his plot had

20. As noted in Appendix B, twenty-seven of the fifty-nine small farmers I interviewed in Cuba came from the CPA and UBPC sectors, and, of these, 81.5 percent were members of a cooperative that had at least some degree of "linking."

21. Bear in mind, however, the caution in interpretation called for by n. 19.

22. See my Appendix B.

been a part of. Yet some from this group came from other sectors of the economy entirely, including one individual who was a retired member of the military.

In spite of the fact that all of these *parceleros* grew crops for their own consumption, they also produced a range of things for the market. Their marketable produce included vegetables, beans, yucca, coffee, dairy products, corn, and plantain. Their marketing patterns varied by crop and by municipality—from selling part of their produce to Acopio, to selling their surplus to those who came by their plots or on the streets of their town.

The *parceleros'* decision about where to market their production was probably influenced by whether they had been incorporated into an already existing CCS, where, in some cases, marketing was conducted collectively. The goal was for neighboring CCSs to bring the *parceleros* into their membership. This would ease the government's responsibilities toward them, while also ensuring that at least part of their production was distributed through official channels. Indeed, by spring 2000, approximately 30 percent of the *parceleros* nationwide had been incorporated into an existing CCS (Association of Small Producers [ANAP], unpublished data, March 31, 2000). Hence, the long-run status of this category of producer remained uncertain.

Nevertheless, when I asked the *parceleros* how they felt about their new-found access to the land, almost all of them spoke exclusively of the advantages that had resulted from the arrangement. These ranged from the economic benefits it afforded them (from no longer having to purchase the food items they were producing and from the sales of their surplus produce), to being able to contribute to the revolution through this means, to feeling more useful as a person. Only one of them spoke of a disadvantage, which was a sense of insecure land tenure. In sum, the majority felt that they had gained a lot by becoming *parceleros*.

In addition to the organizational structure agricultural production assumed, other changes were set in motion in this sector during the 1990s. One of these, as described in Chapter 5, concerned the forms of technology employed in it. Thus, organic production gained an important foothold in the sector when agro-chemicals became unavailable in the early 1990s. In fact, most of the cooperative leaders I met with said that their cooperatives used biological pest control, organic fertilizers, or both of these on at least some of their crops. But, despite the relatively widespread usage of organic inputs, attitudes about them varied. Some said they ought to use them, although they were not doing so at the time (Interview no. 47a, CPA, June 27,

1998), while others felt that their usage was either impractical or backward (Interview no. 1a, CPA, May 28, 1998; and no. 14a, CPA, June 8, 1998). For most, agro-chemicals were still key in their cooperative's production. Therefore, even though organic production experienced a major upsurge in Cuba during the Special Period, it was unclear whether it would remain as prominent if access to agro-chemicals was, once again, unlimited.

Finally, the opening to the world market and community that took place as of 1990 brought cooperative relations with international nongovernmental organizations (NGOs) and foreign governments into Cuban agriculture. While their presence in this sector bore no relation to what it was in either tourism or industry, it was noteworthy (cf. CEPAL/INIE/PNUD 2004, 12). Yet expansion of these relations was somewhat slow, especially outside of the state farm sector where such cooperation was first experimented with on an exclusive basis in the early 1990s. Nonetheless, among the cooperatives included in my sample, collaborative relationships with foreign NGOs and governments were starting to spring up. By mid-1998, when I carried out my survey, 16.9 percent of those interviewed belonged to a cooperative or UBPC that had such relations. In only a few cases did these relations date even as far back as 1994, while most were only several months old.

Cooperative relations typically offered Cuban farmers credit, inputs, technical assistance, and any equipment, such as for irrigation, that was needed. Moreover, the foreign partner took virtually complete responsibility for the international marketing of the resulting crop/product. In some cases, the cooperatives would also receive a foreign exchange bonus (in cash or in-kind) for those products geared for the international market. The products that were the focus of cooperative agreements for my informants were vegetables, citrus crops, and mangos. Undoubtedly, for those cooperatives fortunate enough to attain such an agreement, it greatly ameliorated the limitations commonly plaguing agricultural producers in Cuba at that time.

Key Differences Between the Two Provinces

Historic Patterns

The differences between the Province of Havana and the Province of Santiago de Cuba are striking. Although the former is located next to the City of Havana, by far the largest city in the country, and Santiago de Cuba contains the second-largest city (of the same name as the province), the distinctions between these two provinces go back a long ways in history.

From colonization onward, the Province of Havana's economy was more diversified than that of the Province of Santiago. During the 1500s and 1600s, an assortment of agricultural goods were produced in Havana, including tobacco, sugarcane, coffee, and beef. In addition, subsistence crops were grown there to feed the capital's population. In contrast, during this same period, mining especially for copper was the principal economic activity in Santiago, and it was not until the 1700s that sugar and coffee production also got under way.

In the eighteenth and nineteenth centuries, the infrastructural development of the Province of Havana was noteworthy.[23] Schools and theaters were constructed throughout the province to further its social and cultural development. In the 1850s and 1860s, the telegraph was extended to various parts of Havana. Santiago was not noted for having a comparable sociocultural or infrastructural base at that time.[24]

During the twentieth century, agriculture in both provinces became more diversified. In the Province of Havana, the main agricultural products through the mid-1970s were, in descending order of importance, sugarcane, beef/dairy products, plantain, citrus, and other fruits. Santiago's list of key agricultural products was quite similar, but with slight differences in order of importance: beef/dairy products, sugarcane, coffee, other fruits, and citrus. Both export and domestically oriented agricultural goods were produced in each province. Even though sugarcane cultivation, which is for export, had a greater presence in Havana than in Santiago, plantain—which is domestically oriented—was also among Havana's top three agricultural goods.

An area of more striking differences was that of the two provinces' levels of industrial development. In the mid-1970s, a quarter of the Province of Havana's workforce was employed in industry, compared with 13 percent in Santiago (Editorial Oriente 1977, 45; Editorial Oriente 1978, 49). The former was the second most significant province for industry in the country, after the City of Havana.

Their differing levels of economic development coincided with distinctions between the two provinces in terms of average incomes. Quintana Mendoza (1996, 3–6) documented income differentials between all of the

23. See Editorial Oriente (1978).
24. See Editorial Oriente (1977).

provinces that originated before the start of his study period in 1975. He located Santiago in the category of "backward" provinces and Havana in the category of "forward" provinces with regard to income levels. The coefficient he sets forth (Quintana Mendoza 1996, 3–6) to describe differences in income at the provincial level for the entire country in 1989 was 3.6,[25] a drop from 4.8 in 1975. An obvious trend was under way prior to 1990 for these differentials in income to shrink.

Several attributes of the populations of these two areas of the country may help to explain the historic income differences between them, even in the post-1959 period when the government explicitly sought to reduce regional inequalities.[26] These include the fact that birthrates have traditionally been higher in the Oriente than in western Cuba.[27] This is connected to another difference that existed between the two provinces at this time: the percentage of the population that was working age was somewhat lower in Santiago than in Havana. Both indicators suggest that in the Oriente fewer family members formed part of the workforce than in western Cuba, resulting in a reduced family income in the former region, relative to the latter.

However, while these population characteristics undoubtedly played a role in regional income differentials, other factors probably also contributed to these patterns. For example, the differing levels of development in agriculture between these two provinces led to contrasting yield levels, which, inevitably, affected income levels. There may be parallel examples that can be drawn from comparisons of the two provinces' industrial sectors. In sum, I would argue that a variety of factors generated the historical differences in income, which worsened in the 1990s.

The Post-1990 Period

As a result of the two provinces' economic histories, as well as their physical characteristics, differences between them in the post-1990 period began with what was produced in each. In the Province of Havana, the most important products within the small farmer sector were tubers (especially sweet potatoes and yucca), plantains and bananas, vegetables, and corn. Even between

25. This coefficient represents, in percentages, the magnitude of the difference between the average worker's salary in each province contrasted with the national average.

26. This thesis was presented to me by a senior economist at a socioeconomic research center affiliated with the Cuban government (Interview, May 13, 1999).

27. For example, in 1976, the birthrate in the Province of Santiago was 23.9; in contrast, the Province of Havana's rate was 16.7 (Editorial Oriente 1977, 42, 44; Editorial Oriente 1978, 47).

the two municipalities in Havana there were distinct production patterns: Güira was much more noted for its plantains and yucca, and San Antonio was known for its corn and vegetables. Despite these variations, the crucial contrast was between Havana and Santiago.

In Santiago de Cuba, I found small farmers to be especially strong in cattle raising, particularly dairy farming. This was true for both of the municipalities in which I conducted fieldwork. In Palma Soriano the production of corn was a distant second, followed by coffee and vegetables. In the municipality of Santiago, fruits, particularly mangos, came before dairy farming in significance. Coffee production was third in order of importance. Despite the intraprovincial variation that existed within Santiago de Cuba, it was less remarkable than between the provinces.

Another long-lasting distinction between these two provinces was the level of the forces of production that described them. The Province of Havana was much more developed than Santiago, which was reflected in the production techniques of its small farmers. In the Province of Havana, tractors reigned (100 percent of my survey sample used them), although in recent years, oxen were also reintroduced, and their use was widespread. Irrigation systems that covered most, if not all, of these farmers' cultivated area were the norm (90.3 percent).

In contrast, tractors were much less prevalent in the Province of Santiago. Only 46.4 percent of my survey sample used them; they relied on oxen and manual labor. Likewise, their access to irrigation was notably less strong than in Havana. Only 14.3 percent of those I included in the survey had irrigation. Widespread use of irrigation would probably have ameliorated much of the negative impact of the serious drought of the second half of the 1990s.

An additional difference in production techniques arose after 1990 regarding the use of organic inputs. Their use was more common in Havana (90 percent) than in Santiago (69.2 percent). The reasons for this were probably several: there was greater emphasis placed on the development of organic inputs for the food crops produced in the Province of Havana, which were destined for the capital, in response to the need to ensure food crop production by any means possible; the lower level of infrastructure in Santiago, with its less extensive road systems, would make distribution of some biological pest controls that were relatively "perishable" in nature more difficult; and, given the higher level of productive forces in Havana pre-1990, it was likely that the province had a greater dependence on agro-chemicals, resulting in more need for their replacement during the Special Period.

The patterns of access these two provinces' small farmers had to key agricultural resources also evidenced some interesting differences. For example, the time at which they had gained entrée to work the land differed somewhat between provinces. Among my informants, more had joined the reformed sector of agriculture (i.e., a CPA, UBPC, or had become a *parcelero*) since 1990 in Santiago (60.7 percent) than in Havana (38.7 percent). (See Table 6.1.) This was probably a reflection of the degree to which the Oriente's urban economies were hurt by the economic depression of the 1990s (as was the large-scale migration from the Oriente to the Province of Havana and the capital). Laboring in agriculture became one of the few options open to people in the region.

Another agricultural resource whose pattern of usage varied between these two provinces was credit. The Province of Havana (and Güira in particular) stood out for its greater credit use than the Province of Santiago (51.6 percent and 32 percent of my survey sample, respectively; see Table 6.1). In all likelihood, this was a consequence of the use of more advanced technologies in Havana and the higher costs of production that would result from them. This finding fits well with what Pérez Marín and Muñoz Baños (1992, 40) defined as common features of what they term the "classical model" of agriculture—employment of a high level of technology that is dependent on costly, imported inputs and a close relationship between bank credit and the production process—which had characterized Cuba in general before 1990. It would appear that, at least to some extent, these features still held true in the late 1990s and that the description was more suited to the Province of Havana than to the Province of Santiago de Cuba.

Problems with access to agricultural inputs also appeared to have varied between these two provinces: whereas 29 percent of those who were included in my survey in the Province of Havana said that input shortages had been severe enough to affect their production levels, 53.6 percent of their counterparts in the Province of Santiago had this complaint. A few possible explanations come to mind for this difference. Given the importance of food crop production in the Province of Havana, perhaps it had higher priority in input distribution than Santiago, or perhaps their distribution in Santiago had proved to be more complicated because of the already existing distinctions in these two provinces' infrastructures. Another possibility was that the more widespread use of organic inputs in the Province of Havana had, indeed, reduced the effect of the crisis for that area's farmers.

Within the population of farmers whose production had been affected by input shortages, there were also differences in which sector of production expressed the greatest limitations in this sense. In the Province of Havana, it was exclusively within the CCSs that input shortages had reached the point of negatively affecting production levels. In contrast, it was within the CPA sector that this grievance was most commonly registered in the Province of Santiago.

In addition, farmers in these two provinces differed with regard to what they considered to be the most serious problems faced by Cuba's small farmers at the time of our interview. The issue mentioned by the largest number of my survey participants was input shortages, and their responses were relatively balanced by province. However, the second most commonly listed problem was overwhelmingly mentioned by farmers in Havana, as opposed to those in Santiago: thefts of agricultural produce and livestock (48.4 percent and 14.3 percent, respectively). The proximity of a market with substantial demand and relatively greater purchasing power might well have been responsible for higher levels of theft (if, indeed, figures for theft paralleled perceptions of it).

Another problem farmers in both provinces listed, but somewhat more so in Santiago, was the lack of spare parts for the relatively dated farm equipment that most small farmers had to work with. Those in Havana also spoke of insect infestations, which had a negative effect on production over the previous few years. Santiago's farmers complained of climatic problems and the drought, which had dire consequences for agriculture in the Oriente over the preceding few years. The final problem mentioned by more than a few informants was the U.S. embargo of Cuba.

Last, and most important, Cuban researchers had found that the populations of these two regions had different levels of income (see Quintana Mendoza 1996, 1998).[28] My survey of the two provinces' small farmers revealed a parallel situation to that of the general population. There were remarkable distinctions in their income levels. In the Province of Havana, 80 percent of those interviewed[29] said they earned an average of more than 500 pesos a month, and 40 percent said they earned more than 1,000 pesos a month (see

28. He draws this conclusion from data describing both earnings and savings.

29. Slightly more than 19 percent of my informants in Havana were unwilling or unable to respond to my inquiry about their average monthly income.

Table 6.2). The national average for a monthly salary was 217 pesos in 1998 (Rodríguez 1998, 5), and only approximately 1.5 percent of the working population earned more than 450 pesos a month (Nerey 2003, cited in Togores González 2004, 119). In the Province of Santiago,[30] in contrast, 96.2 percent of the survey participants said that their average monthly income was less than 500 pesos, and 76 percent said it was less than 300 pesos (see Table 6.2). Even if one assumes that people everywhere tend to underestimate their income in surveys, the differences between these two provinces are striking. Being part of the agricultural sector didn't ameliorate the overall tendency for significant income differentials to exist between these regions of the country.

Because of the inevitable shortcomings inherent in accepting without reserve self-reporting on income, I used several additional indicators to gauge how my informants were faring economically. One indicator was their ability to make home improvements. Surprisingly, given acknowledged differences in income, there was relatively little variation between my informants at the provincial level as to how many of them had been able to make some kind of improvement on either their home or farm during the post-1990 period.

Table 6.2 A comparison of four municipalities: The income of small farmers (percentages)

	Province of Havana		Province of Santiago	
Pesos/mo.	Güira (N=15)	San Antonio (N=16)	Palma Soriano (N=15)	Santiago (N=13)
100–200	—	—	60.0	15.4
200–300	13.3	—	26.7	30.8
300–400	13.3	—	—	23.1
400–500	—	6.3	6.7	15.4
500–600	—	18.8	—	7.7
600–700	13.3	18.8	—	—
700–800	—	12.5	—	—
800–900	—	—	—	—
900–1,000	—	—	—	—
1,000–1,500.	13.3	37.5	—	—
1,500–2,000	6.7	—	—	—
2,000–3,000	6.7	—	—	—
No answer	33.3	6.3	6.7	7.7

SOURCE: Author's survey data.

30. Seven percent of my sample in Santiago were unwilling or unable to respond to my inquiry about their income.

In the Province of Havana, 74.2 percent of those included in my survey had done so, although many said that they had not been able to do more than paint their homes (see Table 6.1). In the Province of Santiago, this figure was 67.8 percent. There were, however, differences at the municipal level: they ranged from San Antonio, where 81.3 percent had been able to make an investment of this sort, to Palma Soriano, where only 60 percent had been able to. This range of positions paralleled the extremes in the range of self-declared income levels. Moreover, the starting point for these improvements was quite distinct between the two provinces. That is, most of the small farmers I interviewed in the Province of Havana had homes/farms that were notably better off than those in Santiago.

I also inquired about whether they had been able to purchase any large farm animals or agricultural equipment since 1990. In the Province of Havana, 45.2 percent of them had been able to do so, in contrast to 32.1 percent in the Province of Santiago (see Table 6.1).

Yet another indication of differences in income (broadly defined) between these two regions was my informants' varying access to other benefits that can come from membership in a CPA or UBPC. That is, aside from the direct monetary benefits derived from participating in either of these types of cooperatives, there were additional potential benefits, such as the possibility of purchasing household appliances that were in short supply—at subsidized prices—or having access to a house at the beach or other vacation spot for at least several days a year. Both of these kinds of benefits were more readily available for CPA and UBPC members in the Province of Havana than for those in the Province of Santiago. For example, 60 percent of the CPA/UBPC members included in my survey in Havana had obtained appliances through their cooperative since 1990, in contrast to 33.3 percent in Santiago. Whereas 53.3 percent of the CPA/UBPC members in Havana had taken such a vacation since 1990, only 33.3 percent had been able to in Santiago. Access to these kinds of benefits reflected, above all, the economic status of their cooperatives. But it also represented a component of the income of these farmers.

Finally, the number of people a farm employs also suggests the economic status of the farm. If it is doing well, it will absorb more labor, and if it is doing poorly, members of the farm household may be forced to look for ways to generate income off the farm. Given that this logic holds most true for individual farmers, the data that follow only pertain to CCS members and *parceleros* who farm as household units.

My interviews underlined the striking difference between the two provinces vis-à-vis the absorption of labor on family farms. Small farms in Havana were more able to provide employment for sons, daughters, and other relatives, in addition to the owner, than those in the Province of Santiago. Twice as many informants in Havana than in Santiago were able to provide such employment. The contrast was not as extreme for their farms hiring permanent or temporary workers. But Santiago's small farmers were still less equipped to offer this kind of employment than their counterparts in Havana. Interestingly, it was much more common for the spouse also to be engaged in work on the farm (i.e., for my informant to describe his wife as other than a "housewife") in the Province of Santiago than in Havana. Thus, while small farms in Santiago might occupy the labor time of their owners and their owners' spouses, they were in a much less advantageous position than those in Havana to offer employment to others.

A related issue was that of the career path of these farmers' children. Even though their children did not stray too far from national norms—with their several-decade-long pattern of moving away from employment in agriculture from one generation to the next—those in the Province of Havana remained more tied to the land (either on or off their parents' farm) than those in Santiago (36.1 percent compared with 20.4 percent, respectively). The most isolated of these four municipalities, Palma Soriano, had the lowest percentage of the next generation in agriculturally related employment in this regard (17.9 percent). One possible explanation for this was that, historically speaking, the lower level of income generation of farms in the Province of Santiago did not provide as much of an incentive to stay in this sector of the economy as was the case in Havana.

The Sources of These Regional Income Differences

Multiple indicators from my survey highlight the notable difference in income levels evidenced between the small farmer populations of the Province of Havana and the Province of Santiago de Cuba. My study results coincide with Quintana Mendoza's (1996, 1998) and others' findings that important income differentials were not eliminated in the 1990s. In fact, they argue that the differentials worsened during this period. For example, Quintana Mendoza (1996, 3) documented the trend between 1975 and 1989 toward decreased income differentials between regions. Yet by 1994, the coefficient he calculated to describe these differences was almost as high as it had been in 1975 (it was 4.7 in 1994 as opposed to 4.8 in 1975). My own data do not describe how the income of the small farmers I interviewed changed over time. But the differentials I found were even more dramatic than what Quintana Mendoza and others identified for the population as a whole, suggesting that here, too, there may have been an accentuation of regional inequalities during the 1990s.

What accounted for these differentials among small farmers? I offer several hypotheses to explain this phenomenon. The varying levels of economic development characterizing these two provinces led to quite distinct yield levels between them. Major differences in productivity were evident from the interviews I conducted with cooperative leaders in each province. This already existing difference was compounded in the 1990s by the greater difficulties my informants in Santiago had in obtaining inputs than those in Havana. Finally, the drought of the second half of the 1990s would have exacerbated the situation. The result was that farmers in the Oriente would ultimately have had less produce to sell, which was their principal form of income generation. Moreover, reduced levels of produce implied that they would have had less produce to sell through more lucrative channels once they had met their quotas with Acopio.

In addition, the farmers from the Province of Havana had another advantage over those from Santiago: the group of crops produced by the former included fewer goods that were subject to purchase in their entirety by Acopio or another state agency. That is, once the farmers' consumption needs were met, all of the remaining production of certain goods, including coffee, sugarcane, milk, and meat, had to be sold to a specified state agency. This applied to goods destined for export because state agencies were responsible for most international marketing (except where "cooperative" relationships had been established with foreign NGOs and governments), as well as to key domestically oriented goods that were considered to be basic foodstuffs that had to be distributed through official channels to ensure that consumers had access to them. Although, as mentioned in Chapter 5, coffee and sugarcane producers received special incentives for their produce in the latter 1990s (because they generated foreign exchange earnings) this was not true for milk and meat producers.[31] The particular agencies that purchased domestically oriented goods—Acopio, for example—were known for setting substantially lower prices than what these goods would probably earn in the

31. In 1999 (i.e., a year after my survey was carried out), the price of milk paid to producers was more than doubled (Interview, ANAP official, July 21, 2000).

market. Hence, given the types of farm products those in the Province of Santiago specialized in, they had fewer goods that they could sell their surplus of in the more remunerative Mercados Agropecuarios.

My data put in evidence that slightly more of the farmers included in my survey from the Province of Havana sold their produce in the Mercados Agropecuarios than did those from the Province of Santiago. And slightly more of them from the Province of Havana sold their produce to the state-run agencies, such as Cítricos Ceiba and the Empresa Comercializadora, that were connected to the tourist and export markets and offered higher prices than did those from the Province of Santiago. What this meant was that farmers from the Province of Havana were more able to augment their earnings from produce sold in channels other than Acopio than were farmers from the Province of Santiago.

Finally, because of the income differentials characterizing the overall populations of these two regions, goods sold in the Mercados Agropecuarios in Havana (both in the Province of Havana and, even more important, in the City of Havana) fetched higher prices than those sold in the Mercados in Santiago. In a study of Mercados in Havana, Santiago, and Camagüey, Alvarez (2004, 104) found that prices in the City of Havana were higher for twenty out of twenty-eight key agricultural products in 1999 and 2000 than they were in the other two cities.[32] The price differentials were as high as 300 percent on some items. Therefore, farmers in Santiago were disadvantaged by the lower quantity of goods they had for sale and the prices they were able to sell them at, both of which would negatively impact their income levels.

The Phenomenon of Peasantization

In all likelihood, these regional differences contributed to, but were not entirely responsible for, a larger phenomenon that emerged in the 1990s in Cuba: peasantization. I found evidence of the migratory trend that confronted Cuban planners during the 1990s: people leaving the Oriente and

---

32. Nova González (1995) had documented a similar pattern comparing Havana to a number of other parts of the country in the first couple of years of the markets' existence, as well as for 1999 (Nova González 2004a). See also Díaz Vásquez (2000). This continued to be true well into the following decade (Interviews, researcher, government research center, July 4, 2005; researcher, university research center, July 5, 2005).

moving to the Province (if not the City) of Havana. This was illustrated in the group of people I interviewed from the UBPC sector in the Province of Havana.[33] Those who migrated in the 1990s were not active in agriculture before coming to Havana. Thus, a move from nonagricultural work to agricultural work was taking place across regions.

Some of these UBPC members, in all probability, formed part of what another informant (Interview, government planner, February 15, 2001) described as "directed migrations" that occurred at the beginning of the 1990s. These migrations entailed the organizing of agricultural Contingentes in the Oriente. The Contingentes were sent to work on *granjas*, many of which were then transformed into UBPCs, in the western part of the country. Indeed, some of the UBPC members from the Oriente whom I interviewed on farms in the Province of Havana had been Contingente members. By the time I interviewed them in 1998, their original two-year commitment that incorporation into a Contingente required had passed. Their continued membership in the UBPC indicated their intention to remain in the countryside even as Cuba ascended from the depths of its economic crisis, the effects of which, especially food shortages, had undoubtedly led them to join in the first place.

However, the UBPC sector in the Province of Havana became the site of peasantization for others, as well, who were not recent migrants from the Oriente. Rather, the difficulties of the Special Period and the comparatively higher salary they could earn in an UBPC, along with the food it provided them access to, had made incorporation into an UBPC more attractive to them than the nonagricultural labors they had previously been engaged in.

Yet this process of peasantization was even more apparent in the *parcelero* sector that was established in the 1990s. Some Cuban analysts were skeptical that a genuine and lasting "return to the countryside" had occurred, arguing instead that it was a phenomenon of the early 1990s that had stopped by the middle of that decade (e.g., Interview, two government planners, February 15, 2001; Interview, ANAP official, July 21, 2000). But, data describing the growth of the *parcelero* sector in the latter 1990s contradicts their assertion: whereas in 1996, there were 43,015 *parceleros*, in 1998 (as mentioned in Chapter 5), there were almost 75,000 of them, and by spring 2000, there were

33. However, the overall number of UBPC members who had migrated to the Province of Havana from the Oriente was evenly split between those who had come before 1990 and those who had come later.

almost 97,400.[34] The number of *parceleros* leveled off after that.[35] Nonetheless, even as the economic situation continued to stabilize through the middle of the 2000s, the *parcelero* sector did not experience any reversal in size.

In my own survey, the pool of farmers I interviewed in Santiago contained more *parceleros* than those interviewed in the Province of Havana. This pattern coincided with official data detailing the provincial distribution of *parceleros*: as of spring 2000, there were approximately four times as many people who had received land in this form in the Province of Santiago than in the Province of Havana (ANAP, unpublished data, March 31, 2000). In fact, 12 percent of all *parceleros* were located in the Province of Santiago.[36] This would suggest that the movement of workers from nonagricultural labor to agricultural labor was under way even within the Province of Santiago. The nonagricultural sector in Santiago had been harder hit by the economic crisis/reform than that of the Province of Havana.[37] Consequently, more people who were not previously agriculturalists were pushed into farming than in the Province of Havana.

At the same time, evidence of peasantization during the 1990s could also be found within the CPA sector of agriculture. This was true among the CPAs I visited: 35 percent of the CPA members included in my survey had joined their cooperatives during the 1990s, having worked outside of agriculture before that. A similar dynamic was found by a Cuban researcher who conducted a study of the CPA sector (Interview, February 13, 2001). In her project, more than 20 percent of the CPA members were new, having come from outside of agriculture to join a cooperative. Likewise, in a case study of farming in the municipality of Santo Domingo in the Province of Villa Clara, Sáez (2003) found that new people were flowing into the CPA sector to such an extent that cooperatives had to turn applicants away. Data from several years later would seem to suggest that growth in the CPAs reached a high point in the mid-1990s, after which it began a gradual descent (ONE 2004, table X.4). Yet for all of the recent CPA recruits, higher than average

34. The figure for 1996 is from Martín (1998, 5); that for 1998 is from CEPAL (2000, 315–16); and for 2000, the figure comes from unpublished ANAP data (2000).

35. See further, Pagés (2005).

36. There are fifteen provinces in Cuba (including the City of Havana), so if the number of *parceleros* were to be evenly divided between them, each should have roughly 6.6 percent of the total *parceleros*.

37. A Cuban researcher argued that this was especially true of the measures that formed part of the economic reform and were aimed at reducing excess liquidity (Interview, researcher, economic research center, Havana, May 13, 1999).

incomes and a secure source of food had made the CPAs an attractive alternative for employment.

In addition, it appeared that the influx of new members into the cooperative sector had increased the youthfulness of this population. According to Martín (1998, 10), while "the average age of small farmers was 46 at the end of the 1980s . . . the majority of new members are younger than 40 years of age." Within the pool of those I interviewed who had only recently begun to work in agriculture, and had begun to do so by joining a CPA, most were under forty years old. The relative youthfulness of the new recruits would allow them to develop experience with agricultural work that they had lacked at an age still young enough to take advantage of it. As important, because of their relative youth, it represented an infusion of potentially more energetic people into the agricultural workforce, people who still had a few decades of work ahead of them before retirement.

These various trends point to a shift in Cuba's occupational structure that, albeit not massive, was certainly noteworthy. Official data document it (ONE 2003, table V.1). Between 1981 and 2000, the percentage of private farmers (including *parceleros*, independent *campesinos*, and CCS members) in the labor force rose from 7 to 13 percent, and then grew another approximately 1.5 percent in the following two years. CPA members rose from composing 1 percent to almost 8.5 percent of the labor force during this two-decade period, before dropping to just under 8 percent in 2002.[38] These data confirm a conclusion reached in a study carried out by IPF/FPNU (1996b, 40) that, along with the City of Havana—because of the presence there of the "newly emerging economic" sectors discussed in Chapter 5—rural areas became an "area of attraction" for the population in the 1990s.

Their "attractiveness" in the context of economic crisis and reform can be easily understood when one takes into account the relative advantage they offered in terms of income, as well as the basic sustenance they guaranteed access to. The income data presented earlier highlighted the favorable position small farmers found themselves in compared with the country's average income earner.[39] This advantage grew with time: the income of private farmers (including *parceleros*, CCS members, and independent *campesinos*) rose

38. Martín (1998) and Espina Prieto (2004) also present quantitative data documenting this process of peasantization.

39. Mesa-Lago (2003, 80) locates "private farmers" (one would presume this includes *parceleros* and other unorganized producers, and CCS members) within the upper income earning groups.

424 percent between 1994 and 2000, and that of CPA members rose by 50 percent (calculated from ONE 1998, table IV.1; ONE 2002, table IV.1). These figures for increased income contrasted strongly with those for state employees (27 percent) during the same period. While UBPC members' incomes only grew by 19 percent between 1994 and 2000 (calculated from ONE 1998, table IV.1; ONE 2002, table IV.1), it is essential to reiterate that they, as opposed to workers outside of agriculture, had guaranteed access to food at greatly reduced prices, as well as other amenities. That is, cash earnings were significantly devalued for those who were completely reliant on them for the purchase of food and other indispensable items, given the quite limited supply of goods within official distribution systems and the high prices prevailing outside them. Thus, peasantization represented, in a number of ways, a reasonable alternative in these otherwise hard times.

This peasantization process was not entirely spontaneous. Arias Guevara and Castro Hermidas (1998, 35) assert that, "if in the decade of the 1970s official policies stimulated a process of depeasantization . . . in the current period the state, itself, has created the legal conditions for . . . a situation that has generated an inverse social mobilization: from the city to the countryside." Through its organization of agricultural Contingentes, the formation of UBPCs, the legalization of farmers' markets, and the turning over of land to *parceleros*, the state made peasantization not only a viable but also a comparatively appealing option in the post-1990 period.

At the same time as the state created the material and legal conditions for peasantization, it also engaged in an effort to politically valorize small farming. As resources of all kinds became increasingly scarce, it was these farmers who were heralded as repositories of knowledge about agricultural production and whose methods were seen to offer a way out of the crisis in agriculture, which had such a major effect on the economy. As Powell (2004, 11) states, their representative organization (ANAP) also "gained in stature and autonomy," as these producers came to be considered "integral to a national project." This constituted a remarkable shift from the emphasis on socialized agriculture, with its agricultural workers presented as the most advanced sector of production, that had prevailed from roughly 1965 onward. It was a notable modification as well of the characterization of small farmers (especially those farming on an individual basis) that accompanied the closing of the farmers' markets in 1986, as effectively being a capitalist class that was willing to get rich on the backs of workers. The shift in political valorization reflected the changed conception of this sector from being principally seen as social subjects to also being seen as crucial economic actors in Cuba's reconfigured socialism.

Conclusion

Two important patterns emerged from my study of the impact of economic crisis and reform on Cuba's small farmers. The first was the "peasantization" that was initiated in the 1990s. The second was the maintenance, if not accentuation, of income disparities between these two provinces during the Special Period. Even though these patterns were clearly distinct, they were also interrelated because the second contributed to forming the conditions for the first.

An early indication that a process of peasantization was getting under way took the shape of the agricultural Contingentes that were sent from the Oriente to the western part of Cuba. The Contingentes represented a response to two problems experienced in the Oriente: (1) the growing food crisis and relatively limited options for resolving it within the region and (2) the increasing difficulty of ensuring full employment as the region's industrial sector went into recession. But peasantization also came to be seen as an alternative for an expanding group of people who had formed part of the urban working class in western Cuba. The promise of access to food and a higher income were essential in luring people to work the land. Despite the fact that the promise was less generous within the Oriente, there, too, it came to draw people into agriculture who had not previously seen work in this sector as an attractive option.

Although farming represented a viable option for those from the especially hard-hit Oriente, it was not as remunerative there, as the evidence of substantial income disparities between farmers in the eastern and western parts of the country demonstrated. Nonetheless, even in Cuba's poorer regions, as illustrated by the Province of Santiago, small farmers were no worse off than the average worker in the country. In the Province of Santiago as a whole, roughly half of my informants reported having an income higher than the national average and half reported their income as lower. The self-reported incomes of those interviewed in the municipality of Palma Soriano were significantly lower than those in the municipality of Santiago and, on average, below the national average. Here it is worth repeating, however, that findings from all over the world indicate that people tend to underreport their income. Moreover, the agricultural products that all of the farmers interviewed had access to represented an additional source of income (in the form of savings derived from a reduced need to purchase food)[40] that put them at a distinct advantage over workers in other sectors. This suggests that small farmers in the Province of Santiago were modestly better off than the average salary earner. Thus, it seems reasonable to conclude that Cuba's small farmers were, to varying degrees, better off economically than most wage earners, despite the economic crisis and reform.

Cuba's economic crisis forced that country's policy makers to reshape agricultural policy so that it effectively redressed some of the imbalances that have characterized development policy in many parts of Latin America and that have consistently hurt small farmers and their domestically oriented production. With the shift in agricultural policy that took place in Cuba after 1990, domestically oriented agricultural production achieved a much higher level of priority than it had in the past, and with it, the position of the country's small farmers was fortified.

Intervention by the Cuban government, in the form of a controlled opening to the market, resulted in favorable circumstances for the country's small farmers. The government's decision to reconfigure socialism, thereby leaving intact some elements of the formerly prevailing redistributive hegemony (à la Szelenyi and Kostello 1996), translated into its commitment to move Cuba in the direction of food self-sufficiency. This, in turn, required a prioritization of food production and those engaged in it. While inequality definitely increased overall in Cuba with the market opening of the 1990s, particular sectors of the population—those who played a key role in ensuring the survival of that redistributive hegemony (or reconfigured socialism)—were buffered from the ill effects of that opening. The product of their labors fed back into the society, in the process guaranteeing that most other sectors of the population experienced the market opening to a much lesser extent than they would otherwise have done.

---

40. Even though CPA and UBPC members purchased the goods produced in the "self-sufficiency" section of the farm from their cooperatives, they did so at the price of the cost of production, thereby experiencing a significant savings over what the goods would have cost otherwise.

Conclusion

The policies pursued by the Nicaraguan and Cuban governments toward their respective agricultural sectors after 1990 differed markedly, most especially those polices that focused on food crop production. The outcomes of those policy differences for the farmers involved were also, logically, quite distinct. The position of Nicaragua's food crop producers, who were primarily small farmers, was notably undermined after 1990. In contrast, the small farm sector in Cuba experienced a significant expansion and strengthening during this same period.

These strikingly varied trajectories reflected the larger process of political economic change that each government promoted: in Nicaragua's case, a rapid retreat from socialism; in Cuba, a reconfiguration of socialism. As suggested in Chapter 2, Nicaragua's retreat from socialism, as expressed through its food crop policies, resembled the strategy of the Russian regime toward its agricultural sector as it, too, engaged in a rapid retreat from socialism in the post-1990 period. Cuba's fortification of food crop production and those responsible for it after 1990 likewise resembled the shift in agricultural policy carried out by the Chinese government between 1978 and 1985, as it initiated its own reconfiguration of socialism. In this chapter, I will analyze more closely these resemblances, while also underlining the important differences between countries in each pair. That the results of these distinct approaches to incorporating market relations into previously existing socialist economies and societies were similar in the two paired cases—Nicaragua/Russia and Cuba/China—despite the major differences between the countries in each pair points to the generalizability of my findings.

The commonalities that will be highlighted also have implications for the areas of theoretical debate introduced in Chapter 1. The overarching discussion this study is engaged with concerns the societal effect of expanding market relations, especially as set forth by Karl Polanyi (1944). His analysis of the changes wrought by the spread of market relations during the first era of liberalism resonate with the social and economic consequences of the contemporary era of neoliberalism. Moreover, although describing the shift from feudalism to capitalism, Polanyi's (1944) assertions speak to the more recent transition toward the market experienced by socialist states, too. This study addresses the literature that has emerged to describe this latter process as well—most especially that focused on its social and economic effects for the populations involved. This study examines the transition to the market in Nicaragua, Cuba, Russia, and China through the lens of agriculture, as embodied in the situation of small farmers. Therefore, its findings also intersect with the extensive debate that emerged about the eventual fate of the peasantry in the context of the expansion of capitalism into agriculture. These four cases, particularly in their respective pairs, highlight the contrasting scenarios for the peasantry stemming from this transition process. As was evident, its fate was largely dependent on the vision each government had of its potential role in the economy/society. This vision flowed logically out of these governments' general political economic orientation, whether they sought to move toward a fully capitalist economy or instead were committed to keeping the previous redistributive model hegemonic. Thus, I will also examine the theoretical implications of these cases and their commonalities.

Commonalities/Differences in the Status of Small Farming During the Transition to the Market

Nicaragua/Russia

The last decade of the twentieth century opened with a dramatic shift in the reigning political economic orientation in Nicaragua and Russia. These two formerly socialist regimes were displaced by governments whose principal mission was to bring about a rapid retreat from socialism. The other side of this agenda was to definitively institute a model of capitalism that was wide open to the international market, which was expected to propel their economies "forward."

The key policy mechanism that was relied on in each of these cases to set these two related processes in motion was structural adjustment (SA; see Figure 2). SA was both imposed from without by the international lending community and warmly embraced by each country's new leaders. SA in Russia and Nicaragua had major consequences for their agricultural sectors, particularly that part devoted to food crop production.

A central component of Nicaragua's and Russia's SA programs was the privatization of former state and cooperatively owned properties (see Figure 2). The process encountered a number of obstacles in each of these cases and was, by the end of the 1990s, much closer to completion in the former case than in the latter. By 2000, the previously existing state sector of Nicaragua's economy, which was admittedly much smaller than the Soviet state sector, had largely disappeared. Ownership disputes were still being resolved in various sectors of the economy, including with regard to agricultural land, but

Fig. 2   Similarities within each pathway to the market

Rapid retreat from socialism		Reconfiguring socialism	
Nicaragua	Russia	Cuba	China
Imposition of structural adjustment		Alternative economic reform	
Privatization of land		Downsizing of agricultural production (Commitment to collective ownership)	
Reduced access to agricultural resources:		Increased access to agricultural resources:	
Land less accessible —		More land available	Land available
Drastic reduction of agricultural credit		Credit available	—
Input constraints		Organic inputs available	Inputs more available
		Traditional inputs limited	
Deteriorating terms of trade		Improved terms of trade	
Move toward subsistence production		Push to expand production	
Growing poverty and inequality		Farmers doing relatively well	Farmers doing better (1978–85) Growing inequality and poverty (post-1985)
	Many leaving the land (depeasantization)	Peasantization	—

there was no disputing the existence of a new property regime that was overwhelmingly private. The privatization process in Russia contrasted sharply with that of Nicaragua. Despite several important decrees in the early 1990s, rather than the breakup of large state and collective proper-ties—in the case of agriculture, farms—what occurred was repooling of the individual property certificates that the government had forced to be distrib-uted and the continued presence of a significant area of collective produc-tion. Given that producers were clearly not heeding the call to become the independent capitalists that the architects of SA had envisioned, in 2002 new legislation was enacted that brought the privatization process to its logi-cal end point by permitting the sale of the properties privatized by the de-crees of 1991 and 1993. Although this led to a more active rental market, the real-estate market remained relatively limited in scope even with this legisla-tive stimulus.

Nonetheless, in both countries the dismantling of the state farm sector and the undermining of the cooperative/collective sector created great inse-curity in the countryside. In the Nicaraguan context, privatization coincided with the demobilization of the rival armies that had participated in the Con-tra war of the 1980s, thereby contributing to the violence that frequently exploded in the countryside as this process moved forward (Abu-Lughod 1998; Dye, Butler, Abu-Lughod, Spence with Vickers 1995). Violence was less of an issue with regard to privatization in Russia. But, in both countries, the insecurity and difficulties associated with privatization formed an inaus-picious backdrop for farming in the new era.

Privatization was joined by another central component of SA, the appli-cation of austerity measures, in reducing small farmers' access to the re-sources they needed to produce in Nicaragua and Russia (see Figure 2). That is, privatization ultimately reduced the amount of land under the control of small farmers in Nicaragua, with this being the principal resource they needed. And, retrenchment in government spending greatly limited their access to other agricultural resources. Key among these was credit for ag-ricultural production. In both countries, state bank credit for small farmers, even those held up as models for Russia's new agricultural capitalism, came to be largely nonexistent over the 1990s, and commercial banks overwhelmingly excluded them from their clientele. Those in this sector went from a situa-tion of relative credit abundance in the preceding period to eking out a living with no financial assistance to help them through the long agricultural cycle.

Lack of credit had serious implications for small farmers in Nicaragua and in Russia with regard to yet another resource: agricultural inputs (see Figure 2). In the highly inflationary environment of the late 1980s and 1990 in Nicaragua, the ability to save enough of one's earnings at the end of the harvest not only to ensure the family's survival over the next six to twelve months until the following crop came in but also to purchase the inputs needed for the next production cycle was rare. The same was true in Russia, but high levels of inflation continued to prevail there through 1998, making the starkness of this situation even more lasting. If inflation was not enough, small farmers' purchasing power for inputs in Nicaragua and in Russia was further undercut through SA by the deteriorating prices they received for the agricultural goods they produced (see Figure 2).

The terms of trade crisis functioned in somewhat different ways in these two countries. In Russia, virtually all of agriculture was hurt by the economy's SA. No sector of agriculture was protected when the price of industrial goods used in its production rose disproportionately in relation to producer prices. Food crop production, though, was also hurt by the food imports that flooded into the country as tariffs were slashed. In contrast, given the post-1990 Nicaraguan governments' explicit promotion of export production and their faith in the comparative advantage of importing food rather than producing it locally, it was food crop producers who were hardest hit. Small farmers were most discriminated against, as various mechanisms were put in place to incentivize export production to the exclusion of food crop production. Hence, while agricultural producers across the spectrum suffered from a reduced ability to obtain inputs, it was small farmers whose purchasing power was most affected and whose access to inputs dropped off notably after 1990.

Another difference between these two cases is important to point out. In Nicaragua, the shift to a "free market" was much swifter and more thoroughgoing. This is not to suggest that there were no large players monopolizing the market post-1990, including the inputs market. There were. However, Nicaragua's economy between 1979 and 1990 was of a mixed nature, whereas the Soviet state had, since the Stalinist era, controlled virtually all sectors of the USSR's economy. Therefore, in Nicaragua the transition to a more competitive market had fewer obstacles. The transition in Russia was, instead, characterized by the transformation of many massive state enterprises into similarly massive "corporations." So farmers there also had to contend with huge monopolies as they purchased inputs and sold their produce in a market in which prices were already working against them.

All of these factors together brought about a dramatic drop in production in Russia. As discussed in Chapter 2, even with the privatization decrees of 1991 and 1993, the overwhelming majority of area under cultivation continued to be concentrated on large farms. These were the various types of cooperatives that were formed when those who had been state farmworkers and collective members opted to stick together rather than become independent farmers. Nonetheless, the drop in production was almost exclusively concentrated in the "collectively worked" sections of these cooperatives, where grains were the principal product. On the farms' "personal plots," production expanded. The plots proved crucial for the survival of cooperative members but also for their extended family members who found their purchasing power greatly reduced in the post-1990 era. In essence, production moved more and more in the direction of subsistence (see Figure 2).

In Nicaragua, as mentioned in Chapter 3, food crop production did not contract during the 1990s despite small farmers' lack of credit, the negative terms of trade they confronted, and the food imports that brought down producer prices. But, as also discussed therein, this was largely because the end of the Contra war meant that a notable part of the countryside, which had been uncultivable because of the war, subsequently became cultivable. The population of food crop producers grew in number, and their farmland did likewise with the demobilization of the Contra and many former members of the Nicaraguan armed forces. Thus, the documented increase in food crop production did not reflect a response by farmers to a positive economic environment (i.e., an intensification of production) but rather an extension of areas of cultivation in the central mountain region and further into the agricultural frontier in the search for a means of survival (see Figure 2).[1]

Success at even mere survival became an increasing challenge in both Nicaragua and in Russia. In Nicaragua, rural poverty figures belie data showing expanded corn and bean production (see Figure 2). Small farmers became more impoverished at the same time as alterations in landholding patterns and Gini indices put in black and white the growing income disparities that were readily apparent in the countryside and within the population as a whole. Small farmers in Russia found themselves in similarly dire straits as of the beginning of the 1990s, as the data describing poverty and inequality that was presented in Chapter 2 illustrate (see Figure 2).

---

1. Thus, peasants in both Russia and Nicaragua were engaging in the "subsistence fallback," a phenomenon that provides small-holders with partial autonomy and allows for physical survival in a way that fully landless populations do not have (Bryceson 2000, 312).

Yet the change from the pre-1990 period must certainly have been even more poignant for Russia's small farmers. Before the imposition of SA, the state farms and collectives that farmers belonged to were effectively "total institutions"; they administered to all of their needs from birth until death. But, with the economic streamlining of the 1990s, the new cooperatives were supposed to become nothing more than organizations geared for production. Their social safety net function was to disappear, and it was, indeed, greatly weakened in terms of their ability to provide the social services they once did because of their lack of resources. At the same time, economic hard times put the private purchase of these services beyond the reach of small farmers.

In contrast, Nicaragua's cooperatives and state farms had been, first and foremost, production units during the 1980s. In the areas of the country that were affected by the Contra war, they also had a role in the defense effort, and very limited social services might have been available on some state farms although not at all on cooperatives. Hence, Nicaragua's small farmers had not looked to these institutions for their social safety net—instead relying, and even then not being able to do so to anywhere near the same extent as their counterparts in the USSR, on the public health care and educational systems. Nonetheless, even as comparatively restricted as these services were, their availability was significantly diminished with the government retrenchment of the 1990s.

For Nicaragua's small farmers, SA translated into a notable increase in their marginalization. Their decreasing access to productive resources had seriously negative consequences for their farm-generated income. And, government cutbacks in social services further undermined their standard of living. Small pockets of farmers were able to protect themselves, or even thrive, if they happened to specialize in a product that could make its way into a beneficial niche in the international market, or if they were organized in such a way as to catch the attention of international NGOs that might invest their limited resources in development projects for this sector. But, for most small farmers, the prospects for a dignified existence on their land became much more remote. Their options, then, became to somehow adjust to their increased marginalization by accepting a reduced standard of living or to combine several income-generating strategies at the household level, such as sending one or more family members out in search of employment (to a labor pool that was, itself, shrunk by the SA of the economy), engaging in petty commerce or artisan goods production, or letting a family member leave the country for work in the agricultural, service, or *maquila* sectors elsewhere in Central America or the United States. Yet even the pursuit of these options did not enable all of Nicaragua's small farmers to hold on to their principal agricultural resource—land. Once they lost hold of their land, their only secure tie to a small farmer existence was severed. As I described in Chapter 3, between 1990 and 2001, the percentage of land held by small farmers fell notably, while that held by medium and large farmers increased commensurably. This did not necessarily mean that those who lost their land during this period left agriculture altogether. They could rent land to farm or move onto the agricultural frontier and homestead. But, their hold on farming was markedly diminished. They represented the clearest expression of the marginalization of small farmers brought about by the SA of Nicaragua's economy and its retreat from socialism.

The result of the various economic policies initiated in Russia in 1990/91 was analogous for that country's small farmers. Some analysts argue that a process of "repeasantization" took place there during the 1990s. For example, Kitching (1998b) sees it having resulted from state farmworkers and collective farm members converting to subsistence farmers. Alternatively, in making this case, Burawoy (2000) points out that a large percentage of the country's population (whether urban or rural) had come to depend on personal plot production to meet its consumption needs. However, the overall increase in dependence on the food provided by *dacha* and personal plot production did not mean that the position of small farmers was strengthened—quite the contrary—or that they were even necessarily doing better, relatively speaking, than the rest of the population. They were not. Instead, these processes coincided with the undisputed marginalization of small farmers and the exodus of large numbers of them, especially the youth, from the countryside.

In sum, the policies toward food production pursued by the Nicaraguan and Russian governments after 1990 were similar in a number of senses (see Figure 2), despite the obvious differences between the two countries. Their differences included their size, geopolitical importance, and resource base, as well as the degree to which their economy and society had been socialized before 1990. Moreover, and most interesting, the outcomes of those policies for food crop producers who were, by and large, small farmers were likewise similar. Their imposition of SA, which was designed to propel their economy's and society's rapid retreat from socialism, completely undercut the position of food crop producers, relegating them to a marginalized economic role and a marginalized existence.

Cuba/China

In strong contrast to Nicaragua and Russia, the paths toward the market taken in China between 1978 and 1985 and in post-1990 Cuba led in quite another direction, which had distinct implications for their small farmers. Rather than a retreat from socialism, what each of these latter governments sought was to reconfigure it. Given that in each case the government was acting, at least in part, in response to problematic economic circumstances, some SA-like measures were adopted—such as the partial opening up of their economies to foreign investment. In Cuba, this particular measure was largely driven by the need for a financial infusion, while in China, it was technological innovation that drove the opening. Yet in both cases, the state continued to play a major role in the economy and a redistributive approach to policy making remained hegemonic (see Figure 2).

In the agricultural sector, Cuba's and China's policies were even further distinguished from those pursued in Nicaragua's and Russia's post-1990 trajectories. Cuba and China both engaged in a massive downsizing of agricultural production (see Figure 2). Agricultural production in both countries had been very large scale. This underwent a significant change during each country's move to reconfigure socialism. In Cuba, production was downsized to the cooperative level, as well as within cooperatives to the small-group level. In China, downsizing of production brought control over the production process to the household level. But ownership of the land was not privatized in either country. In Cuba, ownership of the newly formed Basic Units of Cooperative Production (UBPCs) land remained in the hands of the state. Despite the "linking the person with the area" strategy adopted in many CPAs (as well as UBPCs), land ownership remained cooperative. The desire to keep intact a collective approach to production was central. In China, land ownership continued to be held by the community until 2008. In both countries, an underlying concern to avoid the reconcentration of land that often accompanies its privatization was at the core of the land ownership policy.

Another area in which agricultural policy in Cuba and China differed from that characterizing post-1990 Nicaragua and Russia was with regard to small farmers' access to productive resources (see Figure 2). As mentioned in Chapters 5 and 6, access to land increased for small farmers in Cuba during the 1990s. Through the formation of the UBPCs and the allocation of *parcelas* to those willing to work them, farming became an alternative for many people who had never before considered working in this sector of the economy. For others—some of whom became *parceleros*—these policy changes formalized a relationship they had held informally with the state farms they had previously worked for, which allowed for a more long-term investment in their *parcela*. In China, the way in which production was downsized also permitted households to invest in their land, at the same time as it guaranteed them the possibility of expanding or contracting the area they worked as their family size changed.

Yet other resources employed in agriculture also continued to be available or became even more so in this new era for small farmers in Cuba and China. In Cuba, agricultural credit remained within reach of the country's small farmers, if they needed it. Traditional agro-chemicals, diesel fuel, and spare machinery parts became much more difficult to obtain for farming. But, in the allocation of the limited supplies of such traditional inputs as existed, as well as the newly developed organic inputs, small farmers were not discriminated against. If any discrimination occurred, it was by region, with those farmers located on more fertile land and closer to the huge urban population of consumers concentrated in the capital city of Havana appearing to have an advantage. In contrast, given its greater capabilities in this regard, the Chinese government expanded local production of inputs in the 1970s to increase their availability for all farmers. Hence, small farmers' access to this key resource actually grew substantially during this early period of post-Mao reform.

At the same time, the prevailing terms of trade between the producer prices they received for their agricultural goods versus the prices charged for the industrially produced goods they needed for production and consumption were an essential element in determining farmers' access to productive resources. In both Cuba and China, the terms of trade improved notably for small farmers during their respective reform periods (see Figure 2). In Cuba, this took place through the opening of the Mercados Agropecuarios and the new state agencies set up to sell agricultural produce in the international and tourist markets. The government also established incentive schemes that raised prices for some of the goods it purchased through previously existing agencies. Nonetheless, those farmers (regardless of size or organizational form) who produced goods that were not permitted to enter any of these channels or had not had incentivized prices set for their products did not experience these improved circumstances. As the prices of industrialized goods rose, they undoubtedly suffered in the form of lost income and decreased access to such goods. Huge state farms were just as likely to find themselves in this position as small farms. Farm size was not a factor in determining this. Those farmers located in areas within reach of the capital's farmers' markets, where prices for their goods were higher, also had an advantage. But this, too, operated independent of producer size.

In China, terms of trade improved significantly for farmers between 1978 and 1984 through the government's raising of producer prices for all kinds of goods, as well as the opening of farmers' markets. In contrast to Cuba, prices for industrialized goods were held constant in China, thereby benefiting farmers further. But, this situation changed dramatically as of 1985, when the Chinese government shifted its policies concerning prices and quotas. As mentioned in Chapter 2, as of 1985, it allowed producer prices to fall and once again raised procurement quotas, which had been reduced in the late 1970s. Both of these measures meant that farmers found themselves back in the disadvantageous position they had occupied before 1978. However, those farmers located in more remote and less fertile regions had the added disadvantage of being less able to ameliorate these negative conditions by selling part of their produce in the more remunerative urban farmers' markets. So, regional disparities, which had emerged in the first stage of reform, grew.

All of the various agricultural policy initiatives taken in Cuba in the post-1990 period and in China between 1978 and 1985 were designed to foster production increases in agriculture as a whole, but especially in the food crop sector (see Figure 2). Cuba faced a serious food shortage with the fall in trade with the Council of Mutual Economic Assistance (COMECON), which contained the potential to threaten the existence of the regime. Chinese policy makers were concerned about a number of imbalances in the agricultural sector, among them that food crop production was not keeping pace with population growth. This resulted in an increasing need to import food. After 1990, Cuba saw growth in some food crops, such as plantains/bananas and vegetables, and decline in others, such as root crops. Clearly, the far-from-ideal circumstances of the Special Period and climatic problems took their toll on the effort to expand production. In China, food production expanded considerably between 1978 and 1985. This was particularly true for livestock production, but also, to a slightly lesser extent, for crop production. Chinese farmers had the advantages of receiving the many resources that were directed their way between 1978 and 1985 and of not being forced to contend with the many shortages and other hardships that characterized the 1990s for Cuba's farmers.

Nonetheless, Cuba's farmers, like China's farmers during the 1978–85 period, experienced a notable improvement in their level of income (see Figure 2). As mentioned in Chapter 6, Cuba's individual farmers benefited the most in income increases, which were striking, but across the board farmers saw their incomes rise. They also found themselves at an advantage relative to most other workers in the country. Regional disparities in earnings were evident for farmers during this period, following the pattern of the population as a whole. Yet, even given these disparities, farmers in the less advantageous areas were still generally better off in earnings than most workers in the country.

The income of China's farmers also underwent a significant increase between 1978 and 1985. Moreover, their earnings improved relative to urban workers, although they still did not achieve parity with them. These trends were reversed as of 1985, however, when agricultural policy shifted once again and returned farmers to the less favorable position they had occupied earlier. Consequently, the post-1985 period witnessed a growth in rural poverty. In addition, regional income disparities expanded during the post-1985 period, leading to rising inequality between regions.

What was the overall outcome of these various changes for small farmers in Cuba and China? As described in Chapter 6, the Cuban government's efforts to increase food crop production, which included the opening up of access to land in multiple organizational forms and the encouragement of urban unemployed workers to shift to farming, produced a process of peasantization[2] (see Figure 2). While the 1990s did not witness a massive emigration to the countryside, a definite pattern of movement from nonagricultural to agricultural employment was apparent in the recently formed UBPC and

2. At the same time, Cuba shared with Russia a new phenomenon that in the former country was called "urban agriculture." In Cuba, this phenomenon basically consisted of the cultivation of any open spaces existing in urban areas to grow food crops. Urban agriculture was roundly supported by the Cuban government as a means of addressing the food crisis of the 1990s (Companioni et al. 2002; Wright 2005). Although not directly supported by the Russian government, and located in the areas immediately surrounding the country's urban centers rather than within them, urban dwellers took to *dacha* production with a fury in the 1990s (see Zavisca 2003). While the existence of *dachas* long preceded the post-1990 period, it was most especially in this latter period that urban dwellers came to rely on them for *food*, as opposed to as a recreational site for the better off. Thus, after 1990, food products derived from *dacha*—as well as the "personal plot" production of extended family members—became a mainstay in the diet of Russia's urban dwellers.

Yet those who engaged in "urban" and *dacha* agricultural production continued to be urban dwellers and workers. Food crop production in these two contexts was, effectively, a "leisure time" activity rather than a new occupation. Hence, I am distinguishing it from the phenomenon of peasantization that was also under way in Cuba in the 1990s.

parcelero sectors of agriculture. It was also evident in the entry of new members into the CPA sector. And, this pattern continued throughout the 1990s.

In China, the various government initiatives directed at farmers between 1978 and 1985 resulted in a rejuvenation of the agricultural sector, as well as the livelihoods of those employed within it (see Figure 2). As the allocation of labor within farming households gained a new measure of freedom, engagement in sideline production took off. But livestock and crop production also thrived. Whereas the agricultural sector had been somewhat stagnant before 1978, producers responded very favorably to the incentives the reform policies provided to them, and their standards of living benefited accordingly.

In sum, the Cuban and Chinese governments implemented similar policies toward food crop production, and thus their small farm population, as they initiated the reconfiguration of socialism in each of these countries. Despite the clear differences between them, which paralleled those between Nicaragua and Russia in terms of their absolute size, their position on the international geopolitical stage, and their natural resources, they both chose to delimit their opening to the market and to foster agriculture, especially food crop production, and the results of this approach evidenced important likenesses.

Implications for Debates Concerning the "Transition to the Market" and the "Agrarian Question" Within That Transition

The commonalities in the pathways taken toward the market and their social consequences in Nicaragua and Russia and in Cuba and China have a number of implications for the theoretical discussions that have developed about the effect on society of expansions in market relations, especially, but not only, during the "transition to the market" from orthodox socialism, as well as the "agrarian question." Analysis of the effects of this transition on small farmers in each of these countries sheds abundant light on the point of intersection of these several debates.

Theoretical discussion concerning the agrarian question first emerged with regard to Europe, as distinct countries on the continent initiated the transition from feudalism to capitalism. Although the peasantry was freed from serfdom in this process, the question remained as to what would become of it with the expansion of capitalism into the countryside. As described in Chapter 1, Lenin concluded that the latter would generate social differentiation among the peasantry, which would eventually lead to its separation into two classes, a small rural bourgeoisie and a large rural working class.

However, socialism in each one of these four countries interrupted, to varying degrees, the process of capitalist expansion. At least two of the central elements in production—land and labor—became less "commodified" to use the language of Polanyi (1944). Hence, the issue of the future of the peasantry within capitalism was temporarily eclipsed.[3]

The more recent shift toward the market in Nicaragua, Russia, Cuba, and China made the agrarian question relevant once again. As with the initial expansion of capitalism into relations of production, or the emergence of a market economy, as Polanyi calls it (1944), the state in each one of these countries played the key role in paving the way for this process. In Nicaragua and Russia, the socialist state lost credibility, leading to its electoral ouster in the former country and its nonviolent overthrow in the latter, and was replaced by a new state committed to bringing about a full-scale retreat from socialism. In contrast, in Cuba and in China, economic (and, in the latter case, political) crises forced the incumbent regimes to initiate a reconfiguration of their socialist development models. In each case, these processes coincided with the growing importance of international economic relations. Yet Russia's and Nicaragua's states chose one pathway toward the market, and Cuba's and China's states chose quite another.

Those guiding the transition from socialism in Nicaragua and Russia sought a full embrace with capitalism. In both cases, their socialist states had experienced tremendous pressure from external powers during their tenure. And, external actors in the form of multilateral lending agencies largely linked to those external powers, rushed in to "help" define the new policy framework that would propel the retreat from socialism. Together, they chose a pathway leading to unbridled capitalism, rather than one in which a socialist orientation might still have had a role in policy making. The means they employed to achieve both the retreat from socialism and the embrace with capitalism was through a neoliberal restructuring of the economy. As had been true a century earlier when liberalism first gained the day in Europe (see Polyani 1944), the model of SA they imposed conformed closely to that

---

3. This issue was replaced with the question of the future of the peasantry within socialism for the duration of the experiment with socialism in each of these cases.

imposed elsewhere, what came to be called orthodox SA. Thus, it entailed privatization of state and cooperative land, cutbacks in government spending on social services and in support of production, currency devaluations, and liberalization of trade.

Seen in this light, it is not surprising, then, that the results of those policies bore a number of similarities. Polanyi (1944) had argued that the unfettered expansion of the market economy (i.e., capitalism) had resulted in massive human degradation. That degradation took the form of increased economic insecurity and poverty for the laboring classes, as well as profound cultural dislocation, as their old "way of life" was destroyed. When referring specifically to the peasantry, Polanyi (1944) coincided with Lenin in his conclusions concerning their fate—that this sector of the rural population would inevitably disappear as a consequence of its increasing marginalization from the agricultural means of production.

Writing about transitions from socialism, fifty years later, Szelenyi and Kostello (1996) concurred with Polanyi's (1944) pessimistic characterization of the impact of a full engagement with capitalism. Although not focused directly on the peasantry, the former argue that the greater the degree of market expansion within socialist economies, the more inequality will grow.

As we have seen in examining the transitions from socialism embarked on by Nicaragua and Russia, some of these assertions definitely hold true. Small farmers in Nicaragua and Russia experienced increasing marginalization as SA moved forward. Most found their ties to the agricultural means of production weakened through privatization and other measures that made up the SA. And, their livelihoods were accordingly undercut. Poverty levels rose as SA proceeded apace, and the small farmers' rural way of life started to give way as the younger generation migrated in search of employment to urban areas within their respective countries or abroad.

Yet as seen in Chapter 4, the effects of Nicaragua's SA were not entirely uniform. While most of the country's small farmers did experience growing marginalization, not all did. There were small pockets of farmers who stayed afloat, if not thrived. The two factors that were key for their success in this endeavor were (1) producing a good that was doing well in the international market and (2) being a member of an organization that was able to provide at least some of the resources that had previously been available to these farmers through government channels but no longer were. The first of these factors was simply fortuitous for these farmers; their good fortune that what they traditionally produced was blessed with a booming market through the opening up of international trade. Others were far less lucky in the international marketplace, as was the case of the farmers in León who had previously relied on cotton production, when "their" product underwent a "bust," with its devastating implications for their lives and livelihoods. However, a significant number of even the "fortunate" farmers had also opted to organize themselves to safeguard their interests.

The efforts of this group of small farmers, as well as those who fared less well with regard to the terms of trade for the goods they produced, to protect themselves through membership in various types of producers' organizations clearly made a difference for their well-being. These efforts resembled the "societal reaction" that Polanyi (1944) spoke of. Polanyi's societal reaction represented a "countermovement" to the "movement" represented by the expansion of the market economy. It was embodied in distinct social groups' struggles to defend themselves vis-à-vis the ravages of the market economy.

In the nineteenth-century England that Polanyi was writing of, countermovements took multiple forms, as described earlier. In Nicaragua, as well, societal reaction took multiple forms. These ranged from the almost continuous revolts of the early post-1990 years waged by the "recontras" and "recompas,"[4] to the crippling strikes in urban areas engendered by privatization and the rising cost of living caused by government cutbacks and the devaluation of the currency, to the search for nongovernmental avenues to resolve the hardships produced by the new economic regime. Society in Nicaragua did not remain passive in the face of the onslaught brought on by the SA of its economy. The strong history of organizing that had characterized the years of Sandinista governing and, for many, the last years of the war against Somoza, was evident in the immediate and massive nature of the reaction that took place as soon as the SA policies were initiated. Despite the ferocity of the reaction, the SA moved forward. Thus, those who joined together in a more long-lasting way, such as in producer organizations, were ultimately better equipped than most to withstand the ongoing impact of the government's economic policies.

In addition to these more "organized" approaches, individual household members also worked together to expand their income-generating potential

---

4. The *recontras* were groupings of former Contra members who had rearmed after the demobilization of 1990 and were waging war on the Chamorro government. The *recompas* were former Sandinista army members who had also rearmed after their demobilization. Although the first group was more active than the latter, and engaged in its struggle for significantly longer, both groups were ostensibly fighting to ensure that the Chamorro government fulfill the promises it made to them when they demobilized.

beyond what they earned on their own farms. Household members pooled their income from a variety of sources, hiring out as laborers, sending a family member or two to the city or even abroad to work, and so on.[5] As Deere (1990) had found as the transition to capitalism progressed in Northern Peru many years earlier, by pursuing more than one survival strategy a foothold might be kept in the countryside despite deepening marginalization on the farm. Yet Kay (2006, 472) refers to this phenomenon as "de-agrarianization," thereby underlining the decreasing importance of agricultural activities in peasants' livelihoods.

Simply maintaining a foothold in the countryside was far from the ideal of being considered an essential contributor to the new "economy-in-the-making." In fact, Nicaragua's smaller-scale agricultural producers were not seen in the same light by the country's policy makers as those who had promoted liberal policies during the era in England that Polanyi (1944) describes. In this earlier period, the peasantry were desperately needed to form a new industrial labor force, which thereby justified its "separation" from any means of livelihood in the countryside. Struggling for their survival more than a century later, Nicaragua's small farmers were not seen by that country's neoliberal proponents (and their international support organizations) as economic actors who might actually play a productive role in the economy. As in nineteenth-century England, they were perceived as a social sector that was destined to disappear. But, in Nicaragua, while their pain from that process might be mitigated somewhat,[6] their role in the structurally adjusted economy was to sit on the sidelines while economic activity was fostered in other sectors of the society/economy.[7]

Russia's policy makers saw their "traditional" small farmers (i.e., former state farmworkers and collective members) in the same light. These farmers were "holdovers" from the past, who needed to be swept away. Only the new "capitalist farmers" were seen in a positive light, although very little in the

5. Edelman (2008, 248) found this to be true throughout Central America. Among other indicators of this trend was the growing importance of remittances from abroad, which in Nicaragua grew from representing 22.7 percent of export earnings in 1995, to 54.6 percent of them by 2005.

This combination of survival strategies was pursued elsewhere in the Latin American countryside following the implementation of SA. See, for example, Davis (2000), de Janvry, Gordillo, and Sadoulet (1997), and Gravel (2007) on Mexico.

6. Through, for example, temporary measures such as "food for work" programs.

7. Araghi (2008) speaks of this historical difference in noting that in the earlier period there were both push (the Enclosure Movement) *and* pull (the need for factory workers) factors at play, whereas in the contemporary period only push factors exist.

way of agricultural resources were made available to them either, to back up their supposed prioritization. Despite an increased reliance by urban and rural dwellers alike on small-scale production for their sustenance, small farmers found their livelihoods undercut at every turn.

Nonetheless, there too, rural society reacted to the "movement" embodied in the expansion of market relations. That is, there was countermovement in Russia as well. One of the clearest forms it took was the resistance on former state farms and collectives to privatize agricultural resources. Thus, the pooling of land certificates represented a distinct kind of countermovement.[8] More individual approaches to resistance also emerged, however. These included the support provided to extended families in the shape of provisioning from "the personal plot." But, it also included some of the individual and familial survival strategies found in the Nicaraguan case, such as sending family members away to urban areas and abroad in search of an income that could contribute to the well-being of those remaining in the countryside. Many youth, in particular, opted to leave the countryside.

As can be seen, there were both similarities and differences in the forms that countermovement took in Nicaragua and Russia. In his analysis of these dynamics during the earlier "liberal" era, Polanyi (1944), likewise, found the "reactions" of society to diverge between England and the European continent. By way of explanation, he described the distinct course industrialization had taken in each of these two areas and the implications this had for their working classes. It was only logical, then, that the countermovements emerging in the two places would be different. Given the diverse circumstances in which the expansion of market relations—or "movement"—took place in Nicaragua and Russia, it was to be expected that there too, such responses would vary. In spite of multiple expressions of countermovement, in Russia as in Nicaragua, those left behind in the rural areas experienced deepening economic and social marginalization.

In strong contrast, Cuba's and China's economic crises forced their government policy makers to reevaluate their earlier perspective on small farming and those who engaged in it. In both cases, the socialist government had previously sought to move away from small-scale agricultural production—in Cuba through the promotion of large-scale state and collective farming and in China through the latter of these options. Larger-scale production was perceived to be more economically efficient, but individual small farmers

8. I am indebted to Jane Zavisca for this observation.

were also thought to be politically backward. Yet with the reconfiguration of socialism in both countries, small farmers came to be seen as important economic actors. That is, it was assumed that the incentives offered by having a more direct relationship between effort exerted and rewards reaped that was inherent in small-scale farming would lead to much-needed production increases.

Initiatives to strengthen small-scale production were complemented by the liberalization of marketing. This liberalization of markets conformed quite closely to what Szelenyi and Kostello (1996) speak of in terms of the introduction of local markets within redistributive economies. Hence, farmers' markets were opened within these still largely planned economies that continued to be characterized by a strong emphasis on redistribution.

The incentives to engage in small farming bore fruit in both countries, although ostensibly more so in China than in Cuba. In the former country, production and peasant incomes increased dramatically while this economic approach prevailed. In Cuba, a trend toward peasantization emerged alongside production increases in some key food crops, but the effect on production was less than uniform. Despite this, the economic advantages to being a small farmer in Cuba were evident everywhere in the 1990s.

Szelenyi and Kostello (1996) had argued that this controlled approach to introducing markets had the potential to lead to greater equality, as the peasantry would find its standard of living rise relative to urban sectors of the population. This was clearly true for China during the period between 1978 and 1985. But, they also postulated that as more space was allowed for markets, shifting the nature of the economy further in the direction of a socialist mixed one, inequality would increase. This certainly described the situation in post-1985 China, as farmers found many of the supports and protections that had been in place during the earlier period of transition fall away. In Cuba, by the late 1990s the economic model was situated somewhere between these two approaches and the social situation as well. There, however, small farmers found themselves among the better off as inequality increased. They, along with those associated with the newly "emerging economy," had benefited from the expansion of the market, while most of those who had remained employed by the state outside of these two sectors saw their socioeconomic position decline.

Several things distinguished the situation in Cuba from that of post-1985 China. Among these was that Cuba's small farmers were responsible for a valuable good—food—which was in very short supply. As mentioned in

Chapter 6, gaining secure access to food had been the principal motivation for many who participated in the process of peasantization. But that was not all. State support for small-scale farming was crucial in making it an attractive alternative. Whereas state support for small farmers fell away in 1985 in China, it stayed in place in Cuba, even as the emerging economy took hold. While China's policy makers ceased to see agriculture and its producers as a sector that was central to its plans once other sectors took off, this was not true of Cuban policy makers. The latter, instead, sought to continue to increase the availability of food products, at the same time as they hoped to capitalize on agricultural production by linking it up with a few of the emerging sectors (e.g., tourism) as well as by increasing exports. Both of these strategies, in turn, generated much-needed foreign exchange earnings.

This discussion of the extent to which the Cuban government was responsible for fostering small farming brings us back to one of Polanyi's (1944) major points: the expansion of capitalism (or a market economy) is not simply an automatic process. This was true for Cuba. But it was also true of the other three case study countries. In each of these countries, although certainly influenced (if not seriously pressured) by external actors whether they were individual states or international organizations, the state played a key role in determining how much free rein markets would have over the economy. As Polanyi (1944) demonstrated with regard to nineteenth-century (or liberal) England, the expansion of markets was not a "natural" process. The state either blocked or facilitated the process at every step. In so doing, it put in evidence the continued relevance of this institution, contrary to those who argue that its importance has been eclipsed (e.g. McMichael 2006, 2008b).

Likewise, as of 1990 the Nicaraguan and Russian states set their economies on a path toward neoliberal capitalism. Thus, they decided on the future of their small farmers. Reflected in that decision was their vision of the peasantry, as a class destined to disappear. In essence, they concurred with the position that was set forth as early as the late 1800s, that the extinction of the peasantry was inevitable (see Marx 1967; Engels 1977; Lenin 1957).

China's and Cuba's policy makers initially shared with each other a similar vision of the peasantry, as being capable of assisting in the jump-start of the economy. Hence, even as they allowed for a limited opening to the market, they ensured that the needs of this sector would be met. But by 1985, the vision of China's policy makers regarding the peasantry had shifted. And, its allowance of the further expansion of market relations defined the disadvantageous position small farmers would be put in.

Of these four cases, only the Cuban state has held on to its vision of small farmers as important economic actors. There too, the government's vision dictated policies toward the sector, by that means, in large part, deciding the outcome. In Cuba, however, given the distinct vision, government policies have fostered peasantization instead of marginalization.

As Nicaragua, Russia, China, and Cuba initiated their transitions toward the market, those in power in each country had an overarching political economic orientation guiding them as they chose the pathway their countries would take. That orientation contained within it a vision of the role of the peasantry in this change. That vision, then, determined the extent to which the state was willing to let market relations expand at the expense of the well-being of this sector.

It is true that rural society in Nicaragua and Russia responded to this "movement" with "countermovements" to protect itself. And those countermovements were able, at times, to slow down the process of market expansion or ameliorate some of its effects. As a result, pockets within rural society succeeded in staying afloat. Other rural dwellers chose to react more individually (or as a household) by adopting varied survival strategies, including "opting out" through emigration, so that the family might keep a grasp on the land. Regardless of these reactions, the state continued to push the movement onward.

In Cuba and China (between 1978 and 1985), though, the state's reconfiguration of socialism embodied a decision to take a distinct pathway from orthodox socialism. That pathway used, while tightly controlling, the expansion of market relations. In so doing, China's and Cuba's policy makers defined a different future for small farmers, a future in which small farmers would contribute to the economy rather than be superfluous to it. One of the consequences of the decision to define small farmers in this way was to strengthen their economic position within society. During this period in China, the country's small farmers moved closer to parity with other sectors of the population, as inequality in earnings diminished. In Cuba, by the late 1990s, the economic opening had given rise to increased inequality within society. Yet small farmers had been among the winners within this process. Hence, it was within the context of reconfigured socialism, rather than retreat from it, that small farmers thrived. The continued hegemony of a redistributive approach to policy making was crucial in this dynamic because it dictated the need for increased food production to ensure these countries' food self-sufficiency and to ensure that everyone would have access to at least basic sustenance. The emphasis on fulfilling this need directly benefited small farmers.

But why is policy making that benefits small farmers something that states in the Global South, as well as those in the North, should be concerned about? There are several rationales for such concern, even aside from the intrinsic value of seeking to guarantee the well-being of all sectors of society. These begin with the key rationale of the Cuban and Chinese governments—the role of small farmers in food production, which it was critical to increase. More and more states that are dependent on food imports to feed their population, in the recent period of rising food prices worldwide, came to realize just how vulnerable they are. Decreasing this vulnerability would, in all likelihood, improve the position of their small farmers as well.[9] Where this sector of the population is sizable, it can also contribute to greater or lesser political stability, as seen in post-1990 Nicaragua. The rationales also include the desirability of avoiding rural-urban migration where the urban economy cannot absorb their labor—as in most of the Global South. The logical outcome of such migration is teeming shantytowns where underemployment and poverty are rampant. The other option open to those pushed off the land is migration abroad, which affects the surrounding societies and those in the Global North. The health of this economic sector has implications that extend far beyond its rural hamlets.

In their efforts to reconfigure socialism, the Cuban and Chinese states recognized at least some of these implications and designed their policies accordingly. The results of those policies are evidence that small farmers can play an important role in agriculture and that their production can contribute to a country's overall efforts to achieve balanced economic development. They also demonstrate that there is nothing "inevitable" about the status of the peasantry. Rather, its well-being and economic role, like that of other sectors of the population, is largely defined by the actions of the state. The state's actions, in turn, depend on where it situates itself in relation to the movements of the market economy and the countermovements of society.

---

9. In fact, Kay (2006) presents food import substitutionism, or promoting local food production over imports, as a means to reduce rural poverty.

Postscript

Yet another period of change was initiated in Nicaragua and Cuba in the mid-2000s. In Nicaragua, the reelection of the Sandinista National Liberation Front (FSLN) to the presidency in late 2006 and Daniel Ortega's assumption of that position in early 2007 opened the way for change. In Cuba, the serious illness of Fidel Castro, which was made public in mid-2006, and the election of Raul Castro to the presidency in February 2008 signaled this change.

The return of the Sandinistas to power, albeit in the context of a much more divided left than had been the case in the 1980s, created the possibility of a renewed alliance between these two governments. This alliance was rejuvenated under the auspices of a recently formed regional alliance, the Bolivarian Alternative for Latin America (ALBA), which had been established following the election of Hugo Chávez to the presidency of Venezuela in 1998. With the Cold War over and Venezuelan oil reserves as fuel, ALBA offered (at least in theory) the new Sandinista government and that of Raul Castro a distinct context for policy making than what had come before.

In Nicaragua, the new Sandinista government sketched out a set of priorities and means to achieve them that represented a strong contrast with the overall economic—and especially agrarian—policies of the previous sixteen years of neoliberal governing, in the form of the Plan Nacional de Desarrollo Humano, 2008–12 (National Plan for Human Development, 2008–12). Among the elements of contrast were its prioritization of food crop production and its support for small- and medium-sized producers. The former of these priorities was supposed to reduce Nicaragua's dependence on food imports—and increase its food sovereignty (as defined in Chapter 1)—as well as to diminish rural poverty.[1]

How exactly did Daniel Ortega's government propose to go about putting these priorities into practice? Although land redistribution was not in its initial tool kit (GRUN 2008; Interview, Ministry of Agriculture and Forestry, July 14, 2008), every effort was to be made to ensure that small- and medium-sized producers would not be forced to give up any more of their land. The need for foreign investment and participation by the private sector in the economy ruled out taking away land from the landed, either by the state or through land occupations, to redistribute to the landless and the land-poor. But, even within this "ambiance of the market," several things were working in the government's favor: the worldwide food crisis and the discourse of civil society promotion (Interview, Ministry of Agriculture and Forestry, July 14, 2008). Nicaragua was especially hard hit by the food crisis of the mid-2000s, which coincided with a major increase in fuel prices, another imported item. Given its heavy dependence on imported foodstuffs (a situation particularly fostered by the emphasis on exports at the expense of food crop production inherent in the neoliberal economic policies of the 1990s and the first half of the 2000s), the country was vulnerable.[2] This desperate state of affairs pointed to promoting food crop production and those who engaged in it, especially in light of the call for these priorities from within organizations that composed civil society.

The means by which the Sandinista government planned to fortify the position of small- and medium-sized producers included expanding state-supplied credit for this sector. With funds from ALBA, in 2007 the government provided 42 percent more rural credit targeting small producers in particular than had been offered the previous year; and the indications were good that this upward trajectory would continue in 2008 (SPAR 2008, 8). Yet another government program—"Hambre Cero" (Zero Hunger)—distributed a package of goods, including a pregnant cow, a pregnant pig, chickens and a rooster, feed and seeds for feed for them, fruit and other trees, construction materials for a stable, and training in a variety of areas designed

1. This logic coincides with that set forth by Kay (2006).
2. For example, the country imported approximately 40 percent of the rice its population consumed, the price of which soared to new heights in the world market during this period (Dickerson 2008).

to make their production viable.[3] In the first eighteen months of the program, almost 18,500 families had received partial or complete packages (*bonos*) from the government (SPAR 2008, 16).[4] Technical assistance was also to be provided to small farmers by the state once again. In the first year of Sandinista governing (2007), the number of those receiving this service nearly doubled, and this pattern appeared to be continuing into 2008 (SPAR 2008: 11). The government was also working to get ENABAS back into existence, from the shell it had become during the previous sixteen years. Once again the goal was to have its presence in the market influence the prices paid to producers, while also offering consumers grains at subsidized prices.[5]

The food crisis, and the inflation it and rising fuel prices stimulated,[6] underlined the importance of prioritizing food production and those who engaged in it. Edelman (2008, 239–40) notes that the importance of agriculture within the economy, and in terms of export earnings and employment, diminished throughout Central America after 1990. Yet the Sandinistas' agrarian program contained the potential of reversing these trends there, especially given the increased value of coffee and beans, two products consumed domestically and exported on the international market.[7] The well-being and the economic position of small farmers would be improved, and Nicaragua's food dependence would be simultaneously reduced.

Nonetheless, the Sandinistas' efforts in this regard have not been without criticism. Interestingly, it was not the international lending community, which had been so central to the implementation of structural adjustment since 1990, nor the peak business organizations that were the government's chief detractors.[8] This was because the Ortega government has not strayed

3. Two other components of the program were a health and nutrition program for low-income mothers and children and a feeding program for students in preschool and primary school.

4. One commentator pointed to the incomplete "packages" as an indication of the multiple problems characterizing the implementation of this program. Interview, representative of an organization in the Mesa Agropecuaria (a group of organizations concerned with agricultural production), August 4, 2008.

5. Subsidized foodstuffs (rice, beans, and cooking oil) were, indeed, available in poorer neighborhoods and in abundant quantities. Interview, two residents of a poor neighborhood, July 31, 2008). Spoor (1995) writes at length about the market control exerted by ENABAS in the 1980s.

6. Inflation had reached approximately 23 percent by the middle of 2008 (*La Prensa* 2008a).

7. During the 1990s, Nicaragua imported, received donations of, and exported beans. However, by the beginning of the 2000s, the trade balance in beans was tipped decidedly in favor of exports. Throughout the first half of the 2000s, this small-producer crop was Nicaragua's third most important NTAE in terms of foreign exchange earnings. Its principal markets were other Central American nations, Mexico, and the United States.

8. *La Prensa* (2008b); Interview, independent economist, July 29, 2008; and Ministry of Agriculture and Forestry, July 14, 2008. Moreover, the Interamerican Development Bank, which

from agreements that had been made with these institutions in terms of debt repayment, respect for private property, and other elements of macroeconomic policy. Moreover, in 2005, Sandinista legislators in the National Assembly ratified the Dominican Republic–Central American Free Trade Agreement (DR-CAFTA, which includes the Dominican Republic, Central America, and the United States). It took effect in 2006. The implementation of DR-CAFTA effectively locked the country into an even greater opening to the market.

Among those who could expect to be hurt by DR-CAFTA were the dairy farmers of Esquipulas. The major presence of the multinational corporation Parmalat in Nicaragua's milk-processing sector had allowed it to negotiate into the agreement the right to import unlimited quantities of powdered milk (produced by subsidized farmers in the Global North). This promised to undercut the market for nonsubsidized local producers.[9] Hence, the good fortune of Esquipulas's dairy farmers in the 1990s and 2000s would disappear before long, once again highlighting the vicissitudes of the international marketplace and the risks inherent in relying on it. It remained to be seen whether their other saving grace, organizing to defend their interests, might help them to salvage something from what promised to be difficult times ahead.

Many of the organizations that were established in the 1990s to defend small farmers and other sectors of society continued their efforts in the context of a new period of Sandinista governing. And, it was there, within the organizations that formed the country's civil society, that one could find the government's principal critics.[10] Although positions within the pool of critics varied, what united them was their concern with the lack of transparency with which the government was operating. They charged that the resources that were so crucial for the Sandinista government's social and economic

had backed up the International Monetary Fund's and the World Bank's emphasis on SA in the 1990s and the first half of the 2000s, was, by 2008, providing credit to Nicaragua for food crop production (Álvarez and Canales 2008).

   9. Interview, representative of an organization in the Mesa Agropecuario, August 4, 2008. COHA (2008) also speaks of DR-CAFTA's multiple negative effects for Nicaragua's agricultural producers (and those of the Central American region as a whole), including basic grain and dairy farmers.

   10. The Coordinadora Civil is the largest of these groups of critics. It is actually a network of organizations, fourteen of which are territorially based and nine of which are thematically based (Interview, commission member, Coordinadora Civil, July 29, 2008). It was at the forefront of several large demonstrations against the government in the summer of 2008 (e.g., *El Nuevo Diario* 2008b).

programs, which were primarily derived from ALBA funding, were not channeled through normal government coffers (such as the budget of the Ministry of Agriculture and Forestry, for its programs directed at food production). This meant that they were not subject to external (or societal) auditing. Aside from the issue of whether there had been any malfeasance of the funds, and in fact the charge was commonly made that they were being used for political ends,[11] the question arose as to who would then be responsible for their repayment (i.e., with the state eventually assuming them as foreign debt). Given that massive government corruption deprived the country of crucial resources during the previous sixteen years, as well as the incumbent governments of legitimacy, concern about the potential for, or actuality of, it is not surprising.[12]

In sum, although a redistribution of societal resources is not on this Sandinista government's agenda, if it succeeds in carrying out its pro-small-farmer, pro-food-production policies, this will represent a significant change from what prevailed during the previous sixteen years. Will this bring an end to the dramatic levels of poverty and inequality that characterized the country? Without change of a more structural nature, neither of these social patterns will be seriously undercut. But, if carried out with transparency, they contained within them the possibility of offering Nicaragua's peasantry a more dignified existence, the country greater food sovereignty, and more societal consensus about finding a way out of the "unnatural disaster" that these social indicators pointed to. With the resources being provided by ALBA, in particular, Nicaragua had a unique opportunity to move forward on all of these fronts. Only time would tell whether the country's policy makers would seize that opportunity in a fashion that would ensure that those objectives were reached.

Meanwhile, changes were also under way in Nicaragua's ALBA-partner, Cuba. Membership in ALBA brought new possibilities to Cuba as well. As was true for Nicaragua, but starting even earlier on, Cuba was provided with a crucial breathing space through the resources that ALBA made available to it. Aside from the oil lifeline facilitated by its membership in ALBA, since 2005 Cuba's economy had found yet another generator of foreign exchange earnings—the sale of professional services abroad. Those providing these services worked in a variety of fields, including education, agricultural exten-

---

11. Dickerson (2008) and Interview, commission member, Coordinadora Civil, July 29, 2008.
12. This is to say nothing of the rampant corruption that characterized the forty-seven-year-long Somoza dynasty that was ousted from power in 1979.

sion, and most important, medicine. Income from these services replaced international tourism in importance at that point in time.

However, noteworthy changes were also initiated in the agrarian sector of the economy. These were largely implemented under the mandate of Cuba's new president, Raul Castro. They represented a deepening of the pro-small-farmer, pro-food-production position adopted in the 1990s. And, their implementation bespoke of the continuing problem of agricultural shortfalls. In fact, raising food output had been Raul Castro's top priority since he took the helm (Frank 2008).

Production of a number of key food products did expand in the first half of the 2000s (see Table 5.4.1). But, it was not enough to bring down prices in the agricultural markets,[13] nor dramatically improve access to food for the bulk of the population. In fact, in 2005, a greater percentage (71 percent) of the population's daily nutritional intake was imported than in 1985 (Nova González 2006, 285). Yet the situation took a turn for the worse as of 2005. Cuba was hit that year by both a larger than usual number of hurricanes as well as a drought. Production dropped notably, and the recovery over the following several years was unsteady (ONE 2007, table 9.9). It affected all of the country's key food products: roots and tuber production fell by 29 percent between 2004 and 2007, vegetables by 36.4 percent, beans by 26.9 percent, and citrus by 41.5 percent. Production of rice had been on the increase, while still falling short of national demand by some 60 percent, but experienced a serious reversal as of 2003.[14] Although recovering somewhat in 2006 and 2007, it did not achieve its levels of early in the decade.[15] These figures are particularly troubling when one considers that the area cultivated with sugarcane was reduced by 38 percent in 2002 (see Table 5.2), with the closing of almost half of the country's sugar refineries at that time, and that this land was then to be used for food crop production. Moreover, even with the concentration of sugar cultivation on the lands best suited for it, around the country's most efficient refineries, production of this important crop continued to fall. Industrial yields from the cane also failed to experience a noteworthy improvement. The food situation required urgent attention to reduce

13. Nova González (2006, 280) provides price data for between 2001 and 2005. This data show some vacillation in prices over this period, but an almost across-the-board (in terms of products) increase between 2004 and 2005.

14. For production data, see ONE (2007, table 9.9); and for data on the percentage of national rice consumption covered by that production, see Nova González (2006, 260).

15. From its high point in 2003, rice production had fallen by 38.6 percent by 2007 (by which point it had experienced two years of recovery). See further ONE (2007, table 9.9).

growing food imports, which necessitated the expenditure of scarce foreign exchange earnings and increased the country's food vulnerability.

Raul Castro's election to the presidency of Cuba, which made formal his provisional assumption of this position in August 2006, set the stage for the new period of change. The first modification in agricultural policy was announced on March 17, 2008, and entailed the partial liberalization of the input market for farmers. Cuban economists outside the state structure had been pushing for this for some time, as a means of closing the circle of capital.[16] The form this was to take was the opening of state-owned, hard-currency stores with farming supplies. Since the 1960s, farmers have been allocated these goods by the Ministry of Agriculture, once production plans and sales agreements have been made. But ease of access to them has been an issue since 1990. By charging in hard currency for inputs, the state would have greater resources at its disposal for purchases overseas.

The first group of producers to benefit from this initiative would be dairy farmers (in all productive sectors). Clearly, by placing these farmers first in line, the state was demonstrating its desire to tackle the stubborn problem of low milk production that had existed since 1990. Even though it was still unclear how widely this option would be shared, Cuban farmers were optimistic that it would be expanded to include nondairy farmers as well (*Los Angeles Times* 2008).

In another move to facilitate farmers' efforts, steps were taken to decentralize decision making about a wide variety of issues concerning agricultural production to the local level. Until recently, such decisions were made in the national office of the Ministry of Agriculture. This initiative had the potential to greatly ease farming activities, given the importance of local conditions of all types to this kind of production.

Less than a month later, it was announced that unused land on state farms would be made available to those willing to farm it.[17] This represented an expansion of the *parcelero* program, which had been temporarily on hold since the end of the 1990s. In this new round of land distribution, those who already were *parceleros* could receive more land, with their total limit being forty hectares (*El Nuevo Diario* 2008a). Others who would like to become *parceleros* could receive up to thirteen hectares. As before, these *parcelas* were

16. Interview, economic research center, June 29, 2007; and Conference presentation, university researcher, Havana, May 14, 1999.

17. Although announced on April 1 (Spadoni 2008, 12), the decree authorizing this was not actually released until July 18, 2008 (*El Nuevo Diario* 2008a).

granted in usufruct and were not transferable. Yet the leases for them were long enough to encourage their careful usage: for individual farmers, they were for ten years, and for cooperatives, twenty-five years. These leases could be renewed repeatedly.

This new phase of the *parcelero* program was a response to the continuing concern about the cost of importing food to make up for the shortfall in local production. The Cuban government had been spending an estimated $1.9 billion (U.S.) annually on food imports (*El Nuevo Diario* 2008a). This was a high price to pay for a poor country of this size, especially considering its agricultural potential. However, it was particularly troubling because of the large percentage of agricultural land that was lying fallow. A 2008 government report concluded that 50 percent of agricultural land was either unutilized or underutilized (*El Nuevo Diario* 2008a).[18] This represented an increase of 9 percent from five years previously. Clearly, something needed to be done. Making this land available to those willing to farm it seemed logical.

Shortly after the reinitiation of the *parcelero* program was announced, on April 9, it was revealed that the government was developing plans to attract foreign investment into agriculture, especially rice and livestock production (AFP 2008). Foreign "cooperation" has existed in the agricultural sector since, at least, the early to mid-1990s.[19] But it would seem that the avenues for this investment to take may be more diverse in the future, including providing financial backing for agriculture. Although in making this announcement the Minister of Foreign Investment implied that the government's deliberations in this regard were open to any and all possibilities, it remained to be seen whether foreign investors would be able to purchase agricultural land. Given the country's pre-1959 history of foreign companies controlling much of the sugar industry, foreign ownership of land has been completely prohibited until now in an effort to keep land available for Cuban farmers.

These various measures represented a deepening of the process of change initiated in 1990, geared toward addressing the issues that were still inhibiting the expansion of food production. The process of incorporating elements of the market into agricultural production and commercialization in Cuba has been, and continues to be, a gradual one. Obviously, the state is anxious

18. See also ONE (2007, table 9.3).
19. Alvarez (2004) describes foreign investment in agriculture prior to this time.

for this to be a controlled effort that will allow it to maintain its central, redistributive role in the economy and society. While providing agricultural producers with even more resources—most especially land—or avenues for obtaining them, it seeks to better the lot of those engaging in farming, to increase the access of consumers to food, and to improve the country's food security and macroeconomic balances. Hence, although the market has clearly gained space within the Cuban economy since 1990, and even further since early 2008, redistribution of resources has accompanied it, thereby ensuring the inclusion of small farmers in the economy and demonstrating the state's vision of them as crucial actors in Cuba's "reconfigured" socialism.

A NOTE ON THE USE OF THE TERMS
SMALL FARMER AND *PEASANT*

The focus of this study is small farmers. This sector of the population is, and has been, referred to differently in distinct countries and distinct eras. For example, in Nicaragua, the terms *peasant* and *small farmer* have been, over time, and are currently, used virtually interchangeably. This despite the fact that the former word, at times, denotes cultural attributes and relations of dependency with regard to a landlord class, in addition to the core concept of small-scale agricultural producer. In contrast, the term *small farmer* is a more delimited concept, referring specifically to their scale of production.

In Cuba, the term *small farmer* is used much more commonly, although the term *peasant* is still used, at times, to refer to those engaged in small-scale production. There, though, the concept of peasant is largely devoid of the connotation of dependency, which has its origins in feudal relations in the countryside. The term *small farmer* is used inclusively in Cuba to cover individual small-scale producers as well as those who are members of the variety of cooperative relations that exist there, as witnessed by their organization in the National Association of Small Producers (ANAP).

At the point in time when I conducted the fieldwork for this study in Cuba, members of the recently formed UBPCs, most of whom had previously been workers on the state farms that these cooperatives were located on, were not organized under the auspices of ANAP. Instead, they continued to be represented organizationally by the union they had belonged to on those state farms, the Cuban Labor Federation (CTC). Nonetheless, given their new position as members of cooperatives who ostensibly have a status similar to those of the already existing Agricultural Production Cooperatives (CPAs), I opted to include them in the category of "small producer," and therefore in my study.

Moreover, as indicated in note 1 of Chapter 6, were all of the land within the CPAs and Basic Units of Cooperative Production (UBPCs) to be divided between their members, the area held by the latter would qualify them as small-scale producers. Their inclusion is not meant to imply that they have had, or currently have, a lived experience identical to that of individual small farmers. Rather, given that the focus of this study is on agricultural production and the livelihoods of those who engage in it, the more cultural characteristics typically associated with "peasants," or individual small-scale producers—and the potential lack of these among certain kinds of cooperative members—are less relevant.

Russia represents a large contrast from either of the first two countries. Before the revolution, its rural small-holders fit squarely within the comprehensive definition of the peasantry and were referred to as such. From 1917 to about 1970, however, they were, over time, transformed into state farmworkers and members of collectives. Because of the massive size of these state farms and collectives, the birth-to-death social service coverage provided to their workers and members, respectively, and the internal work relations, those who labored on these estates came to resemble "workers" much more than "small farmers."

With the change in regime that occurred in 1990/91, though, private plot production on the former state farms and collectives grew in importance, while the area of collective production on those same farms was weakened substantially. As described in the Conclusion, this has led some scholars to even describe this process as one of "repeasantization." This was, indeed, a significant shift in status as at least a few generations had grown up and lived their lives, effectively, as workers, rather than as peasants, during the intervening decades. Thus, the concept of "small farmer" became more relevant in Russia in the 1990s than it had been during the mid- and latter-Soviet periods. It is now within reason to include these producers in a comparative discussion of the situation of small farmers with my other three cases.

Finally, in China, collectivization drives during the 1940s and 1950s had led to the incorporation of that country's peasantry, with all that the term implies, into massive collective farms. Yet the gap in time between collectivization and decollectivization that occurred within that country's process of reform meant that, unlike the situation in Russia, some of those who became "small farmers" after 1978 were young enough to have been peasants before the revolution. And, the reversion to individual household production after 1978, as well as the common usage of the term *peasantry* in China, justify their inclusion in my "small farmer" category.

All of this is to note that I am cognizant of the historical and conceptual distinctions. But, the more recent shifts in production relations in each of these countries have led me to feel confident of the appropriateness of categorizing the agricultural producers under study in these four countries as small farmers.

At the same time, I would like to acknowledge a recent debate about the appropriateness, or lack thereof, of the term *peasant* in the current historical period. Kearney (1996) makes the case that this concept is an invention of anthropology that, among other things, is meant to denote a distinction from "modern." He posits that, in the contemporary world in which rural small-holders engage in a variety of nonfarm employment and also at times migration, including internationally, the distinction implied in the term is no longer relevant. Yet, as Edelman (1999) notes for the Costa Rican case, Nicaragua's small producers describe themselves as *campesinos* (peasants), and many of Cuba's small producers still speak of themselves as *guajiros* (or peasants). Moreover, this study is not focused on these farmers' cultural orientation, nor their "identity" per se. Hence, I have opted to rely principally on the term *small farmer* but am not unduly concerned about its correctness where I use the term *peasant*.

APPENDIX B

MY RESEARCH METHODOLOGY

As described in the text, a survey of small farmers was carried out in four rural municipalities in both Nicaragua and Cuba. However, the strategies employed for constructing the survey sample differed in the two cases, given the varying circumstances characterizing the organization of agricultural production in Nicaragua and in Cuba in the 1990s.

In Nicaragua, a snowball sampling procedure was adopted for the survey's construction. This strategy was chosen because there was no comprehensive list of Nicaragua's small farmers (or of the small farmers in each of these municipalities), thereby making the construction of a random sample of the population beyond the possibilities of this study. In each municipality, the entrée for beginning the "snowball" was slightly different.

In Esquipulas and San Dionisio, my research assistant and I began with a list of the small farmers who were interviewed by the research team of a Nicaraguan nongovernmental organization that had conducted a survey addressing "development needs" in the Department of Matagalpa. So as not to duplicate its sample, we only drew a few names from their lists for each of the municipalities and then asked these individuals for more suggestions of whom we might speak with. This process was repeated until we had succeeded in interviewing fifteen small farmers from each of these municipalities. In selecting those who would compose the sample, we made sure to cover multiple rural zones within each municipality, so that our data would reflect the situation throughout the municipalities, rather than in just one or two communities. However, a number of zones were inaccessible because of the activities of armed bands (the *armados*) or because of their distance from roads, which would have entailed several days of back-and-forth travel by foot.

In Malpaisillo and Santa Rosa del Peñon, our entrée was provided by representatives from a local development project, who put us in contact with a few individuals in each rural hamlet. These individuals were the starting point for our snowball. Here, too, we made sure to cover a number of different rural zones. However, because of Santa Rosa's lack of infrastructure, the sample there was limited to those small farmers whose land was located within reach of roads; within an easy hike (e.g., several miles) of a road; within a mule ride that was no more than three hours in each direction; or to those who had come into the municipal capital and were willing to be interviewed there. The broadening out of the snowball was repeated until we had interviewed fifteen small farmers in each municipality.

In Cuba, as mentioned in the "Introduction," I interviewed farmers from all of the nonstate sectors of agriculture (i.e., members of UBPCs, CPAs, CCSs, and *parceleros*). Before interviewing the members selected for inclusion in my sample, in each UBPC, CPA, and CCS, I interviewed one or more members of the Junta Directiva to obtain general information about it as well as entrée to its members. These totaled twenty-one "preliminary" interviews.

The cooperatives included in the survey were selected from the pool of cooperatives in each municipality that met the basic criteria prescribed for this study: their emphasis was on domestically oriented production. Beyond that criteria, in some cases, selection was based on my prior experience with them, and in others, distinct factors such as proximity (given the problem of transportation between locations) and willingness to have me visit them came into play. Once at each cooperative, the individual informants were chosen on a similar basis. The *parceleros* interviewed were selected from a list of *parceleros* in the municipality on the basis of proximity and willingness to be interviewed. The breakdown of informants included in my survey was the following:

Provincia de la Habana

Güira de Melena	San Antonio de los Baños*
5—CPA members	5—CPA members
6—CCS members	8—CCS members
2—UBPC members	3—UBPC members
2—*Parceleros*	

Provincia de Santiago de Cuba

Palma Soriano**	Santiago de Cuba***
5—CPA members	5—CPA members
6—CCS members	8—CCS members
2—UBPC members	
2—*Parceleros*	

* San Antonio de los Baños' *parceleros* were not accessible for interview, given that they were not under the auspices of the organizations that I had access to there (ANAP and the municipality). I opted to include two additional CCS members in their stead.
** Two of the CCS members interviewed in Palma Soriano were *parceleros*.
*** All of Santiago's *parceleros* were affiliated with CCSs. Thus, two CCS members I included were *parceleros*. I was unable to interview any UBPC members in Santiago.

Abbassi, Jennifer. 1997. The role of the 1990s food markets in the decentralization of Cuban agriculture. *Cuban Studies* 27:21–39.

Abu-Lughod, Deena. 1998. Land rights and rural violence in postwar Nicaragua: Circuitous routes to national reconciliation. Paper presented at the XXI conference of the Latin American Studies Association (LASA), September 24–26, Chicago, IL.

Acevedo Vogl, Adolfo. 1993. *Nicaragua y el FMI: El pozo sin fondo del ajuste.* Managua: Latino Editores.

———. 1998. *Economía política y desarrollo sostenible.* Managua: BITECSA.

———. ca. 2004. *Impactos potenciales del Tratado de Libre Comercio Centroamérican Estados Unidos en el sector agrícola y la pobreza rural en Nicaragua.* Managua: Ediciones Educativas, Diseño e Impresiones S. A. (EDISA).

AFP (Agence France Presse). 2008. Cuba mulls more foreign investment in farming sector. *Agence France Presse,* April 9.

Akram-Lodhi, A. Haroon, Saturnino M. Borras, Jr., Cristóbal Kay, and Terry McKinley. 2007. Neoliberal globalization, land and poverty: Implications for public action. In *Land, poverty, and livelihoods in an era of globalization,* ed. A. Haroon Akram-Lodhi, Saturnino M. Borras, Jr., and Cristóbal Kay, 383–98. New York: Routledge.

Akram-Lodhi, A. Haroon, and Cristóbal Kay, eds. 2008. *Peasants and globalization: Political economy, rural transformation, and the agrarian question.* New York: Routledge.

Akram-Lodhi, A. Haroon, Cristóbal Kay, and Saturnino M. Borras, Jr. 2008. The political economy of land and the agrarian question in an era of neoliberal globalization. In Akram-Lodhi and Kay 2008, 214–38.

Allina-Pisano, Jessica. 2002. Reorganization and its discontents: A case study in Voronezh Oblast. In O'Brien and Wegren 2002, 298–324.

Altimir, Oscar. 1994. Cambios en la desigualdad y la pobreza en América Latina. *El Trimestre Económico* LXI 241 (1): 85–133.

Alvarez, José. 2004. *Cuba's agricultural sector.* Gainesville: University Press of Florida.

Álvarez, Wendy, and Gisella Canales. 2008. Crédito BID para comida y energía. *La Prensa,* August 4.

Álvarez González, Elena. 1995. Caracteristicas de la apertura externa Cubana. Paper presented at the XIX conference of LASA, Washington, D.C., September 28–30.

Amador, Freddy, and Gerardo Ribbink. 1993. El mercado de tierra en Nicaragua y el sector reformado. *Revista de Economía Agrícola* 6:3–15.

AMUNIC. 1997a. *Esquipulas.* Departamento de Matagalpa.

———. 1997b. *Larreynaga.* Departamento de León.

———. 1997c. *San Dionisio.* Departamento de Matagalpa.

———. 1997d. *Santa Rosa del Peñon.* Departamento de León.

Análisis Total. 1995. *La dinámica del mercado de productos lacteos.* Tomo I. Consultancy Report. Managua.

ANEC/CEEC/F.E./MINAGRI/MINCIN/ANAP/ONAT/ONE/MEP/INIE (Asociación Nacional de Economistas de Cuba/Centro de Estudios de la Economía Cubana/Facultad de Economía, Universidad de La Habana/Oficina Nacional de Estadísticas/Ministerio de Agricultura/Ministerio de Comercio Interior/ Asociación Nacional de Agricultores Pequeños/ONAT-Ministerio de Finanzas y Precios/Ministerio de Economía y Planificación/Instituto Nacional de Investigaciones Económicas). 2000. Cinco Años de Mercado Agropecuario. Unpublished manuscript.

Araghi, Farshad. 2008. The invisible hand and the visible foot: Peasants, dispossession, and globalization. In Akram-Lodhi and Kay 2008, 111–47.

Arana, Mario J., and Juan F. Rocha. 1998. Efecto de las políticas macroeconómicas y sociales sobre la pobreza en el caso de Nicaragua. In *Política macroeconómica y pobreza en América Latina y el Caribe*, ed. Enrique Ganuza, Lance Taylor, and Samuel Morley, 613–67. Madrid: PNUD/Ediciones Mundi-Prensa.

Arias Guevara, María de los A., and Nelsa Castro Hermidas. 1998. Un enfoque socioclasista hacia el interior del movimiento cooperativo. In *Cooperativismo rural y participación social*, ed. Niurka Pérez Rojas, Ernel González Mastrapa, Miriam García Aguiar, 23–40. Havana: Universidad de La Habana.

Assies, Willem. 2008. Land tenure and tenure rights in Mexico: An overview. *Journal of Agrarian Change* 8 (1): 33–63.

Averhoff Casamayor, Alberto. 1996. Las relaciones de dirección Empresa-UBPC: Situación actual y expectativas. In Colectivo de Autores 1996, 137–51.

Barberia, Lorena. 2004. Remittances to Cuba: An evaluation of Cuban and U.S. government policy measures. In Domínguez, Pérez Villanueva, and Barberia 2004, 353–412.

Barham, Bradford, Mary Clark, Elizabeth Katz, and Rachel Schurman. 1992. Nontraditional agricultural exports in Latin America. *Latin American Research Review* 27 (2): 43–82.

Baumeister, Eduardo, and Edgar Fernández. n.d. Análisis de la tenencia de la tierra en Nicaragua a partir del Censo Agropecuario 2001. Mimeographed.

Bebbington, Anthony. 1998. Sustaining the Andes? Social capital and policies for rural regeneration in Bolivia. *Mountain Research and Development* 18:173–81.

———. 1999. Capitals and capabilities: A framework for analyzing peasant viability, rural livelihoods, and poverty. *World Development* 27 (12): 2021–44.

Bebbington, Anthony, and John Farrington. 1993. Governments, NGOs, and agricultural development: Perspectives on changing inter-organisational relationships. *Journal of Development Studies* 29 (2): 199–219.

Belisario, Antonio. 2007a. The Chilean agrarian transformation: Agrarian reform and capitalist "partial" counter-agrarian reform, 1964–1980. Pt. 1: Reformism, socialism and free market neoliberalism. *Journal of Agrarian Change* 7 (1): 1–34.

———. 2007b. The Chilean agrarian transformation: Agrarian reform and capitalist "partial" counter-agrarian reform, 1964–1980. Pt. 2: CORA, Post-1980 outcomes and the emerging agrarian class structure. *Journal of Agrarian Change* 7 (2): 145–82.

Benjamin, Medea, Joseph Collins, and Michael Scott. 1984. *No free lunch: Food and revolution in Cuba today*. San Francisco: Institute for Food and Development Policy.

Berstein, Henry. 2008. Agrarian questions from transition to globalization. In Akram-Lodhi and Kay 2008, 239–61.

BCN (Banco Central de Nicaragua). 1998. *Indicadores Económicos* IV, 8 (August). Managua: BCN.

———. 1999. *Informe Anual, 1999*. Managua: BCN.

———. 2000. *Indicadores Económicos* VI, 7 (August). Managua: BCN.

———. 2003. *Indicadores Económicos* X, 11 (December). Managua: BCN.

———. 2007. *Indicadores Económicos* (September). Managua: BCN.

Bogdanovskii, Vladimir. 2005. Agricultural employment in Russia. *Comparative Economic Studies* 47:141–53.

Borras, Saturnino M., Jr., 2008. La Vía Campesina and its global campaign for agrarian reform. *Journal of Agrarian Change* 8 (2/3): 258–89.

Boucher, Stephen R., Bradford L. Barham, and Michael R. Carter. 2005. The impact of "market friendly" reforms on credit and land markets in Honduras and Nicaragua. *World Development* 33 (1): 107–28.

Bramall, Chris. 1993. The role of decollectivisation in China's agricultural miracle, 1978–1990. *Journal of Peasant Studies* 20 (2): 271–95.

———. 2000. *Sources of Chinese economic growth, 1978–1996*. Oxford: Oxford University Press.

Broegard, Rikke J. 2000. Land titles, tenure security and land improvements: A case study from Carazo, Nicaragua. M.A. thesis, University of Copenhagen.

Brundenius, Claes. 1984. *Revolutionary Cuba: The challenge of economic growth with equity*. Boulder, CO: Westview Press.

Bryceson, Deborah Fahy. 2000. Disappearing peasantries? Rural labour redundancy in the neo-liberal era and beyond. In *Disappearing peasantries? Rural labour in Africa, Asia and Latin America*, ed. Deborah Bryceson, Cristóbal Kay, and Jos Mooij, 299–326. London: Intermediate Technology Publications.

Bulmer-Thomas, Victor, ed. 1996. *The new economic model in Latin America and its impact on income distribution and poverty*. London: Macmillan and the Institute of Latin American Studies, University of London.

Burawoy, Michael. 2000. Transition without transformation: Russia's involutionary road to capitalism. Paper presented to the University of Chicago Anthropology of Europe Workshop and the Culture, History, and Social Theory Workshop, "Socializing knowledge: Transformation and continuity in post-socialist cultural forms," Chicago, IL, March 10.

Burchardt, Hans-Jürgen. 2000. *La última reforma agraria del siglo: La agricultura Cubana entre el cambio y el estancamiento*. Caracas: Editorial Nueva Sociedad.

Bu Wong, Ángel, Pablo Fernández, Armando Nova, Anicia García, and Aída Atienza. 1995. Las UBPC y su necesario perfeccionamiento. *Cuba: Investigación Económica* 2 (2): 15–42.

Cáceres, Sinforiano. 2003. *Lo agrario y los T.L.C.: Caja de pandora en el movimiento rural*. Managua: Anama.

Cajina Loáisiga, Ariel. 1993. *Nicaragua: Producción y comercialización de productos lacteos*. Managua: Ruta II (Unidad Regional Técnica Nacional).

Cárdenas Toledo, Nora. Forthcoming. *El campo cubano en vispera del 40 aniversario de la reforma agraria*. Havana: Editorial de Ciencias Sociales.

Carranza Valdés, Julio. 1992. Cuba: Los retos de la economía. *Cuadernos de Nuestra América* 19:131–58.

Carranza Valdés, Julio, Luis Gutiérrez Urdaneta, and Pedro Monreal González. 1995. *Cuba: La reestructuración de la economía. Una propuesta para el debate*. Havana: Editorial Ciencias Sociales.

Carter, Michael R., and Bradford L. Barham. 1996. Level playing fields and laissez-faire: Postliberal development strategy in inegalitarian agrarian economies. *World Development* 24 (7): 1133–49.

Carter, Michael R., Bradford L. Barham, and Dina Mesbah. 1996. Agro-export booms and the rural poor in Chile, Guatemala, and Paraguay. *Latin American Research Review* 31 (1): 33–65.

Centeno, Miguel Angel, and Mauricio Font. 1997. *Toward a new Cuba: Legacies of a revolution.* Boulder, CO: Lynne Rienner Publishers.

CEPAL (Comisión Económica para América Latina, Naciones Unidas). 1996. *Información básica del sector agropecuario, Subregión Norte de América Latina y el Caribe, 1980–1994.* Mexico City: CEPAL.

———. 1997a. *La economía Cubana: Reformas estructurales y desempeño en los noventa.* Mexico City: Fondo de Cultura Económica.

———. 1997b. *Series macroeconómicas del Istmo Centroamericano: 1950–1996.* Mexico City: CEPAL.

———. 2000. *La economía Cubana: Reformas estructurales y desempeño en los noventa.* Mexico City: CEPAL/Agencia Sueca de Cooperación Internacional para el Desarrollo (ASDI)/ Fondo de Cultura Económica.

———. 2001. *Información básica del sector agropecuario, Subregión Norte de América Latina y el Caribe, 1980–2000.* Mexico City: CEPAL.

———. 2007. *Social panorama of Latin America 2007.* LC/G.2351-P/I (November).

CEPAL/INIE/PNUD (CEPAL/INIE/Programa de las Naciones Unidas para el Desarrollo). 2004. *Política social y reformas estructurales: Cuba a principios del siglo XXI.* LC/MEX/G.7LC/L.2091. Mexico City: CEPAL.

Chang, Gene H. 2006. Decomposition of China's rising income inequality: Is the rural-urban income gap solely responsible? In *China's rural economy after WTO: Problems and strategies,* ed. Shunfeng Song and Aimin Chen, 134–40. Burlington: Ashgate Publishing.

Chen, Xiwen. 2006. Conflicts and problems facing China's current rural reform and development. In Dong, Song, and Zhang 2006, 33–42.

CIPRES (Centro para la Promoción, la Investigación y el Desarrollo Rural y Social). 2007. *Pequeños y medianos productores agropecuarios: Soberanía alimentaria y desarrollo agroindustrial.* Managua: CIPRES.

Clemens, Harry. 1993. Competividad de Nicaragua en la producción agrícola. Serie CIES/ESECA 94.1, UNAN, Managua, April.

COHA (Council on Hemispheric Affairs). 2008. Dealing with a bad deal: Two years of DR-CAFTA in Central America, November 17.

Colectivo de Autores. 1996. *UBPC: Desarrollo rural y participación.* Havana: Universidad de La Habana.

Companioni, Nelso, Yanet Ojeda Hernández, Egidio Páez, and Catherine Murphy. 2002. The growth of urban agriculture. In Funes, García, Bourque, Pérez, and Rosset 2002, 220–36.

CONAGRO/Banco Mundial. 1993. Caracterización económica, León-Chinandega. Informe 14. Managua, July.

Conroy, Michael E. 1987. Economic aggression as an instrument of low intensity warfare. In *Reagan versus the Sandinistas,* ed. Thomas W. Walker, 57–79. Boulder, CO: Westview.

Conroy, Michael E., Douglas Murray, and Peter Rosset. 1996. *A cautionary tale: Failed U.S. development policy in Central America.* Boulder, CO: Lynne Rienner.

Davis, Benjamin. 2000. The adjustment strategies of Mexican ejidatarios in the face of neoliberal reform. *CEPAL Review* 72:99–118.

Davis, Benjamin, Calogero Carletto, and Jaya Sil. 1997. Los hogares agropecuarios en Nicaragua: Un análisis de tipología. Proyecto GCP/RLA/115/ITA, FAO/UN, November.

Davis, Benjamin, and Marco Stampini. 2002. Pathways out of (and into) poverty in Nicaragua. Background paper prepared for the World Bank Nicaragua Poverty Assessment, May 30.

Deere, Carmen Diana. 1990. *Household and class relations: Peasants and landlords in Northern Peru.* Berkeley and Los Angeles: University of California Press.

———. 1991. Cuba's struggle for self-sufficiency. *Monthly Review* 43: 55–73.

———. 1994. Implicaciones agrícolas del comercio cubano. Declaration before the Subcommittee on Agriculture and Hunger outside of the U.S., Committee on Agriculture, U. S. Congress.

———. 2000. Toward a reconstruction of Cuba's twentieth century agrarian transformation: Peasantisation, depeasantisation, and repeasantisation. In Bryceson, Kay, and Jos Mooj 2000, 139–58.

Deere, Carmen Diana, and Mieke Meurs. 1992. Markets, Markets Everywhere? Understanding the Cuban Anomaly. *World Development* 20 (6): 825–39.

de Groot, Jan P. 1994. Reforma agraria en Nicaragua: Una actualización. In *Ajuste estructural y economía campesina,* ed. Jan P. de Groot and Max Spoor, 97–122. Managua: Escuela de Economía Agrícola, UNAN.

de Janvry, Alain. 1981. *The agrarian question and reformism in Latin America.* Baltimore: Johns Hopkins University Press.

de Janvry, Alain, G. Gordillo, and Elisabeth Sadoulet. 1997. *Mexico's second agrarian reform: Household and community responses.* Transformation of Rural Mexico Series No. 1. Ejido Reform Research Project. San Diego: Center for U.S.–Mexican Studies, University of California.

de Janvry, Alain, and Elisabeth Sadoulet. 1989. Investment strategies to combat rural poverty: A proposal for Latin America. *World Development* 17 (8): 1203–21.

Díaz Vásquez, Julio A. 2000. Los mercados agropecuarios en Cuba. In *La larga marcha desde el Período Especial hacia la normalidad—Un balance de la transformación Cubana,* ed. Jurgen Bahr and Sonke Widderich, 67–85. Kiel: Im Selbstverlag des Geographischen Instituts der Universitat Kiel.

Dickerson, Marla. 2008. Belts tightening in Nicaragua. *Los Angeles Times,* May 6.

Dijkstra, Geske. 2000. Structural adjustment and poverty in Nicaragua. Paper presented at the XXII conference of LASA, Miami, FL, March 16–18.

Dilla, Haroldo. 1999. The virtues and misfortunes of civil society. *NACLA* 32 (5): 30–36.

Domínguez, Jorge I. 2004. Cuba's economic transition: Successes, deficiencies, and challenges. In Domínguez, Pérez Villanueva, and Barberia 2004, 17–47.

Domínguez, Jorge I., Omar Everleny Pérez Villanueva, and Lorena Barberia, eds. 2004. *The cuban economy at the start of the twenty-first century.* Cambridge, MA: David Rockefeller Center for Latin American Studies, Harvard University.

Dong, Xiao-Yuan, Shunfeng Song, and Xiaobo Zhang, eds. 2006. *China's agricultural development: Challenges and prospects.* Burlington: Ashgate Publishing.

Dye, David R., Judy Butler, Deena Abu-Lughod, Jack Spence with George Vickers. 1995. Contesting everything, winning nothing: The search for consensus in Nicaragua, 1990–1995. Cambridge: Hemisphere Initiatives.

Eberlin, Richard. 1998. Comparative advantage of food crops under structural adjustment in Nicaragua. In *Agrarian policies in Central America*, ed. Wim Pelupessy and Ruerd Ruben, 44–75. New York: St. Martin's Press.

Eckstein, Susan. 1997. The limits of socialism in a capitalist world economy: Cuba Since the collapse of the Soviet bloc. In Centeno and Font 1997, 135–50.

———. 2004. Transnational networks and norms, remittances, and the transformation of Cuba. In Domínguez, Pérez Villanueva, and Barberia 2004, 319–51.

ECLA (Economic Commission for Latin America—UN). 1950. *The Economic development of Latin America and its principal problems.* Lake Success, NY: UN Department of Economic Affairs.

Economics Press Service. 2007. *En la encrucijada de la economía Cubana, enfoque especial.* Havana: Economic Press Service.

Edelman, Marc. 1999. *Peasants against globalization: Rural social movements in Costa Rica.* Stanford: Stanford University Press.

———. 2008. Transnational organizing in agrarian Central America: History, challenges, prospects. *Journal of Agrarian Change* 8 (2/3): 229–57.

Editorial Oriente. 1977. *Provincia: Santiago de Cuba, cuna de la revolución.* Santiago de Cuba: Editorial Oriente.

———. 1978. *Provincia: La Habana.* Santiago de Cuba: Editorial Oriente.

El Nuevo Diario. 2001. Marcha es por sobreviviencia. *El Nuevo Diario,* April 22.

———. 2008a. Cuba entrega tierras ociosas a campesinos. *El Nuevo Diario.* July 19.

———. 2008b. Miles de nicas exigen que Ortega Renuncia. *El Nuevo Diario.* July 19.

Emmanuel, Arghiri. 1972. *Unequal exchange.* New York: Modern Reader.

Engels, Frederick. 1977. The peasant question in France and Germany. In *Karl Marx and Frederick Engels: Selected Works.* Vol. 3. Moscow: Progress Publishers.

Enríquez, Laura J. 1991. *Harvesting change: Labor and agrarian reform in Nicaragua, 1979–1990.* Chapel Hill: University of North Carolina Press.

———. 1994. *The question of food security in Cuban socialism.* Berkeley: International and Area Studies, University of California at Berkeley.

———. 1997. *Agrarian reform and class consciousness in Nicaragua.* Gainesville: University Press of Florida.

———. 2000. Cuba's new agricultural revolution: The transformation of food crop production in contemporary Cuba. *Development Report.* Oakland, CA: Food First—Institute for Food and Development Policy.

———. 2007. A new economic model? Linkages between local industry and tourism in post-1990 Cuba. Prepared for the XXVII International Congress of LASA, Montreal, Canada, September 5–8.

Enríquez, Laura J., and Rose J. Spalding. 1987. Banking systems and revolutionary change: The politics of agricultural credit in Nicaragua. In Spalding 1987, 105–25.

Envío. 1993. Medidas económicas: Reactivación solidaria? *Envío* 134:11–19.

———. 1994. Decolectivización: Reforma agraria "desde abajo." *Envío* 13 (154): 17–23.

Escuela de Sociología-UCA. 1987. Elementos para la caracterización política-ideológica de la base contrarrevolucionaria de la V Región: Estudio de zonas de guerra en Chontales y Zelaya. Unpublished report, Managua.

Espina Prieto, Mayra Paula. n.d. Transición y dinámica de los procesos socioestructurales. CIPS, Mimeographed.

———. 2004. Social effects of economic adjustment: equality, inequality, and trends toward greater complexity in Cuban society. In Domínguez et al., 209–43.

Espinoza, María Auxiliadora. 1994. Experiencias de organización de la comercialización de granos por pequeños productores: El caso del Banco de Granos de San Dionisio, Matagalpa. In *Mercados y granos básicos en Nicaragua*, ed. Harry Clemens, Duty Greene, and Max Spoor, 281–99. Managua: ESECA-UNAN/Programa Agrícola (CONAGRO)/BID/PNUD.

Estay R., Jaime E. 1993. Introducción: Economía y reforma económica en Cuba: Una aproximación general. In *Economía y reforma económica en Cuba*, ed. Dietmar Dirmoser and Jaime Estay R., 15–49. Caracas: Nueva Sociedad.

Evans, Trevor. 1995. Ajuste estructural y sector público en Nicaragua. In *La transformación neoliberal del sector público: Ajuste estructural y sector público en Centroamérica y El Caribe*, ed. Trevor Evans, 179–261. Managua: Latino Editors.

Fan, Shenggen. 2006. Rural development strategy in western China under the WTO. In Dong, Song, and Zhang, 78–97.

Fan, Shenggen, and Connie Chan-Kang. 2007. Regional inequality in China: Scope, sources, and strategies to reduce it. In Spoor, Heerink, and Qu 2007, 49–69.

Ferriol Muruaga, Ángela. 1998. Pobreza en condiciones de reforma económica: El reto a la equidad en Cuba. *Cuba: Investigación Económica* 4 (1): 1–38.

———. 2000. Apertura externa, mercado laboral y política social. *Cuba: Investigación Económica* 6 (1): 23–54.

———. 2003. Acercamientos al estudio de la pobreza en Cuba. Paper presented at the XXIV conference of LASA, Dallas, TX, March 27–29.

Fewsmith, Joseph. 1994. *Dilemmas of reform in China: Political conflict and economic debate*. Armonk, NY: M. E. Sharpe.

FIDEG. 1997. Situación económica: Febrero 1997 (La Propiedad y La Cobra). *El Observador Económico* 62:5–23.

Figueroa Albelo, Victor. 1996. El nuevo modelo agrario en Cuba bajo los marcos de la reforma económica. In Colectivo de Autores 1996, 1–45.

Flores Cruz, Selmira, and Ner Artola. 2004. El sector lacteo en Nicaragua: Un vistazo desde la perspectiva de genero. *Encuentro* 36 (67): 150–71.

Forster, Nancy. 1982. The revolutionary transformation of the Cuban countryside. *UFSI Reports*, 26.

Frank, Marc. 2008. Cuba moves to decentralize state-run agriculture. *Reuters*, March 24.

Friedmann, Harriet. 1978. World market, state, and family farm: Social bases of household production in the era of wage labor. *Comparative Studies in Society and History* 20 (4): 545–86.

FUNDENIC/INDES. 1997. Estrategia para el Desarrollo Rural Sostenible: Departamento de Matagalpa (Lineamientos Estratégicos). Unpublished document, Managua.

Funes, Fernando, Luis García, Martin Bourque, Nilda Pérez, and Peter Rosset, eds. 2002. *Sustainable agriculture and resistance: Transforming food production in Cuba*. Oakland, CA: Food First Books.

Funes Monzote, Fernando. 2002. The organic farming movement in Cuba. In Funes, García, Bourque, Pérez, and Rosset 2002, 1–26.

Gambold Miller, Liesl L., and Patrick Heady. 2003. Cooperation, power, and community economy and ideology in the Russian countryside. In *The postsocialist agrarian question*, ed. Chris Hann and the "Property Relations Group," 257–92. Munster: LIT.

Ganuza, Enrique, Lance Taylor, and Samuel Morley, ed. 1995. *Política macroeconómica y pobreza en América Latina y el Caribe.* Madrid: PNUD/Ediciones Mundi-Prensa.

García Álvarez, Anicia. 2004. Sustitución de importaciones de alimentos en Cuba: necesidad o posibilidad? In *Reflecciones sobre economía Cubana,* ed. Omar Everleny Pérez Villanueva, 195–251. Havana: Editorial de Ciencias Sociales.

Gariazzo, Alicia. 1984. El café en Nicaragua: Los pequeños productores de Matagalpa y Carazo. *Cuadernos de Pensamiento Propio,* Serie Avances Dos (December).

Getz, Christina. 2003. Transnational linkages, social capital and sustainable livelihood security: Organic agriculture in Baja, California. Ph.D. diss., University of California, Berkeley.

Goldin, Liliana. 1993. Economic restructuring and new forms of market participation in rural Latin America. In *Politics, social change, and economic restructuring in Latin America,* ed. William C. Smith and Roberto Patricio Korzeniewicz, 105–22. Miami, FL: North-South Center Press.

Gómez, Sergio, and Jorge Echenique. 1988. *La agricultura Chilena: Las dos caras de la modernización.* Santiago: Facultad Latinoamericana de Ciencias Sociales y Agraria.

González, David. 2001. A coffee crisis' devastating domino effect in Nicaragua: Workers go hungry as crop's prices fall. *New York Times,* August 29.

———. 2002. Cuba's bittersweet move to trim its sugar crop. *New York Times,* October 9.

Granma. 1991. Información a la población sobre medidas adicionales con motivo de la escasez de combustibles y otras importaciones. *Granma,* December 20.

———. 1995. Vinculación al area no puede significar creación de minifundios: Dijó Orlando Lugo en encuentro de cooperativistas en Pinar del Rio. *Granma,* August 14.

Gravel, Nathalie. 2007. Mexican smallholders adrift: The urgent need for a new social contract in rural Mexico. *Journal of Latin American Geography* 6 (2): 77–90.

GRUN (Gobierno de Reconciliación y Unidad Nacional). 2008. *Plan nacional de desarrollo humano, 2008–2012.* Managua: GRUN.

Gunn Clissold, Gillian. 1997. Cuban-U.S. relations and the process of transition: Possible consequences of covert agendas. In Centeno and Font 1997, 73–89.

Gwynne, Robert N., and Cristóbal Kay. 1996. Agrarian change and the democratic transition in Chile: An introduction. *Bulletin of Latin American Research* 16 (1): 3–10.

Haggard, Stephen, and Robert Kaufman, ed. 1992. *The politics of economic adjustment.* Princeton: Princeton University Press.

Hamilton, Nora. 1983. *The limits of state autonomy: Post-revolutionary Mexico.* Princeton: Princeton University Press.

Hartford, Kathleen. 1985. Socialist agriculture is dead; long live socialist agriculture! Organizational transformation in Rural China. In Perry and Wong 1985, 31–61.

Heerink, Nico, Marijke Kuiper, and Xiaoping Shi. 2007. Farm household responses to China's new rural income support policy. In Spoor, Heerink, and Qu, 147–60.

Herrold, Melinda. 1993. Cranes and conflicts: NGO programs to improve people-park relations in Russia and China. Ph.D. diss., University of California, Berkeley.

Horton, Lynn. 1998. *Peasants in arms: War and peace in the mountains of Nicaragua, 1979–1994.* Athens: Ohio University Center for International Studies.

Huang, Jikun, and Scott Rozelle. 2003. The impact of trade liberalization on China's agriculture and rural economy. *SAIS Review* 23 (1): 115–31.

Huber, Evelyne, and Fred Solt. 2004. Successes and failures of neoliberalism. *Latin American Research Review* 39 (October): 150–64.

Humphrey, Caroline. 1998. *Marx went away—but Karl stayed behind.* Ann Arbor: University of Michigan Press.

IICA (Instituto Interamericano de Cooperación para la Agricultura). 2003. *Estudio de la cadena de comercialización de la leche.* Managua: IICA.

INEC (Instituto Nacional de Estadísticas y Censos). 1997. *Población: Municipios, Volumen IV. VII Censo de Población y III de Vivienda, 1995.* Managua.

Ioffe, Grigory, Tatyana Nefedova, and Ilya Zaslavsky. 2006. *The end of the peasantry? The disintegration of rural Russia.* Pittsburgh: University of Pittsburgh Press.

IPF/FPNU (Instituto de Planificación Física/Fondo de Población de las Naciones Unidas). 1996a. La distribución de la población en Cuba y su relación con la producción en el ámbito rural. Proyecto Cub/93/P02, December.

———. 1996b. Recomendaciones para una política de distribución espacial de la población. Informe Final. Proyecto Cub/93/P02, December.

———. 1996c. Resultados de la Encuesta Nacional de Migraciones Internas segun niveles del sistema de asentamientos: El caso de la Ciudad de la Habana, December.

IRAM (Institut de Recherches et d'Applications des Metodes de Developpement). 2000a. Estudios sobre la tenencia de la tierra. Pt. I: Marco legal institucional, September.

———. 2000b. Estudios sobre la tenencia de la tierra. Pt II. Inseguridad y conflictos en torno a la propiedad rural en la Planicie del Pacifico: El caso de Malpaisillo y El Jicaral. Case study in "Inseguridad de la tenencia y resolución de conflictos bajo distintos regimenes de derechos sobre la tierra," September.

Irvin, George. 1983. Nicaragua: Establishing the state as centre of accumulation. *Cambridge Journal of Economics* 7 (2): 125–39.

Jonakin, Jon. 1996. The impact of structural adjustment and property rights conflicts on Nicaraguan agrarian reform beneficiaries. *World Development* 24 (7): 1179–91.

Jonakin, Jon, and Laura J. Enríquez. 1999. The non-traditional financial sector in Nicaragua: A response to rural credit market exclusion. *Development Policy Review* 17 (2): 141–69.

Kautsky, Karl. 1976. *The agrarian question.* Winchester: Zwan Publications.

Kay, Cristóbal. 1994. Exclusionary and uneven development in rural Latin America. Paper presented at the XVIII conference of LASA, Atlanta, GA, March 10–12.

———. 1997. Globalization, peasant agriculture, and reconversion. *Bulletin of Latin American Research* 16 (1): 11–24.

———. 2002. Chile's neoliberal agrarian transformation and the peasantry. *Journal of Agrarian Change* 2 (4): 464–501.

———. 2006. Rural poverty and development strategies in Latin America. *Journal of Agrarian Change* 6 (4): 455–508.

Kazakbaev, R. Kh. 2006. Attitudes of the young people of Bashkortostan toward life in the countryside. *Russian Education and Society* 48 (9): 70–79.

Kearney, Michael. 1996. *Reconceptualizing the peasantry: Anthropology in global perspective.* Boulder, CO: Westview Press.

Kelliher, Daniel. 1992. *Peasant power in China: The era of rural reform, 1979–1989.* New Haven: Yale University Press.

Kitching, Gaven. 1998a. The development of agrarian capitalism in Russia, 1991–97: Some observations from fieldwork. *Journal of Peasant Studies* 25 (3): 1–30.

————. 1998b. The revenge of the peasant? The collapse of large-scale Russian agriculture and the role of the peasant "private plot" in that collapse, 1991–97. *Journal of Peasant Studies* 26 (1): 43–81.

Kornbluh, Peter. 1987. *The price of intervention: Reagan's wars against the Sandinistas.* Washington, D.C.: Institute for Policy Studies.

Korzeniewicz, Roberto Patricio, and William C. Smith. 2000. Poverty, inequality, and growth in Latin America: Searching for the high road to globalization. *Latin American Research Review* 35 (3): 7–54.

Laird, Roy D. 1997. Kolkhozy, the Russian achilles heel: Failed Agrarian Reform. *Europe-Asia Studies* 49 (3): 469–78.

La Prensa. 1999. Exportaciones de lacteos suspendidas. *La Prensa,* February 11.

————. 2001. Niños sostienen hogares en mina La India. *La Prensa,* May 16.

————. 2008a. Inflación alta. *La Prensa,* July 22.

————. 2008b. Nicaragua y FMI se entienden sin problemas. *La Prensa,* July 28.

La Tribuna. 1996. No tradicionales estan dinámicos. *La Tribuna,* July 12.

Lehmann, David. 1984. Smallholding agriculture in revolutionary Cuba: A case of underexploitation. *Development and Change* 16:251–70.

Lenin, V. I. 1957. *The development of capitalism in Russia.* Moscow: Institute of Marxism-Leninism.

Lerman, Zvi. 2002. The impact of land reform on the rural population. In O'Brien and Wegren 2002, 42–67.

Lewis, Jessa. 2002. Agrarian change and privatization of *Ejido* land in northern Mexico. *Journal of Agrarian Change* 2 (3): 401–419.

López, Mario, and Max Spoor. 1993. Cambios estructurales en el mercado de granos básicos en Nicaragua. Working paper series CIES/ESECA 93.1, UNAN, Managua.

Los Angeles Times. 2008. Cuba to aid farmers. *Los Angeles Times,* March 18.

Loxley, John. 1984. *Debt and disorder: External financing for development.* Boulder, CO: Westview Press.

Lu, Mai, and Calla Wiemer. 2005. An end to China's agricultural tax. *China: An International Journal* 3 (2): 320–30.

MAG-FOR (Ministerio de Agricultura—Forestal). 1998. *Boletín Trimestral* 1 (July).

————. 1999. *Boletín Trimestral* 2 (May).

Mann, Susan Archer. 1990. *Agrarian capitalism in theory and practice.* Chapel Hill: University of North Carolina Press.

Marquetti Nodarse, Hiram, and Anicia García Capote. 2002. Cuba's model of industrial growth: Current problems and perspectives. In Monreal 2002a, 69–95.

Martín, Lucy. 1998. Campesinado y Reforma. Unpublished manuscript, CIPS (Centro de Investigaciones Psicológicas y Sociológicas), November.

Marx, Karl. 1967. *Capital.* Vol. 1. Moscow: Progress.

Matus L., Javier. 1994. Monitoreo al Mercado de Tierras: Sintesis (Departamento de Rivas). Preparación al Proyecto ALA 93/56, April 1994.

McMichael, Philip. 2006. Peasant prospects in the neoliberal age. *New Political Economy* 11 (3): 407–18.

————. 2008a. Food sovereignty, social reproduction and the agrarian question. In Akram-Lodhi and Kay 2008, 288–312.

————. 2008b. Peasants make their own history, but not just as they please . . . *Journal of Agrarian Change* 8 (2/3): 205–28.

Merlet, Michel, Carlo Fedele, Julio Cesar Quintero, Javier Matus Lazo, and Denis

Pommier. 1993. Nicaragua: Programa de apoyo al fortalecimiento de la situación de derecho y al despegue económico en el campo. Informe de la misión de identificación. NI: EURAGRI, October.

Mesa-Lazo, Carmelo. 2003. *Economía y bienestar social en Cuba a comienzos del siglo XXI.* Madrid: Editorial Colibrí.

Monreal, Pedro. 1999. Sea changes: The new cuban economy. *NACLA* 32 (4): 21–29.

———, ed. 2002a. *Development prospects in Cuba: An agenda in the making.* London: Institute of Latin American Studies.

———. 2002b. Export substitution re-industrialization in Cuba: Development strategies revisited. In Monreal 2002a, 9–29.

Murdoch, William W. 1980. *The poverty of nations: The political economy of hunger and population.* Baltimore: Johns Hopkins University Press.

Murray, Warwick E. 1997. Competitive global fruit export markets: Marketing intermediaries and impacts on small-scale growers in Chile. *Bulletin of Latin American Research* 16 (1): 43–55.

———. 1998. The globalisation of fruit, neoliberalism and the question of sustainability: Lessons from Chile. *European Journal of Development Research* 10 (1): 201–27.

Nakano, Y. 1992. Impacto de los programas de ajuste y estabilización sobre los pobres rurales: El caso de Brasil. In *Ajuste macroeconómico y pobreza rural en América Latina,* ed. R. E. Trejos, 139–203. San José: Instituto Interamericano de Cooperación para la Agricultura (IICA).

Nee, Victor. 1989. A theory of market transition: From redistribution to markets in state socialism. *American Sociological Review* 54:663–81.

———. 1996. The emergence of market society: Changing mechanisms of stratification in China. *American Journal of Sociology* 101 (4): 908–49.

Neira Cuadra, Oscar. 1996. *ESAF: Condicionalidad y deuda.* Managua: Ediciones CRIES.

New York Times. 2004. Cuba: The buck stops. *New York Times,* November 9.

Nickolsky, Sergei. 1998. The treadmill of socialist reforms and the failures of post-Communist "revolutions" in Russian agriculture. In *Privatizing the land: Rural political economy in post-communist societies,* ed. Ivan Szelenyi, 191–213. London: Routledge.

Nolan, Peter. 1988. *The political economy of collective farms.* Boulder, CO: Westview Press.

———. 1995. *China's rise, Russia's fall: Politics, economics, and planning in the transition from Stalinism.* New York: St. Martin's Press.

———. 1996/1997. China's rise, Russia's fall. *Journal of Peasant Studies* 24 (1/2): 226–50.

Nova González, Armando. 1995. Mercado Agropecuario: Factores que limitan la oferta. *Cuba: Investigación Económica* 1 (3): 63–72.

———. 1998. Las nuevas relaciones de producción en la agricultura. Paper presented at the XXI conference of LASA, Chicago, IL, September 24–26.

———. 2003. La UBPC y el cooperativismo en la agricultura Cubana, 1993–2001. Paper presented at the XXIV conference of LASA, Dallas, TX, March 27–29.

———. 2004a. Comportamiento del mercado interno de alimentos agropecuarios 2004. Unpublished manuscript.

———. 2004b. El cooperativismo: Linea de desarrollo en la agricultura Cubana, 1993–2003. Paper prepared for the XXV conference of LASA, Las Vegas, NV, October 7–9.

————. 2006. *La agricultura en Cuba: Evolución y trayectoria (1959–2005)*. Havana: Editorial de Ciencias Sociales.

Nove, Alec. 1986. *The Soviet economic system*. Boston: Allen and Unwin.

O'Brien, David, and Stephen Wegren, eds. 2002. *Rural reform in Post-Soviet Russia*. Washington, D.C.: Woodrow Wilson Center Press/Johns Hopkins University Press.

OECD (Organization for Economic Co-Operation and Development). 2005. *OECD review of agricultural policies: China*. Paris: OECD.

Oi, Jean C. 1989. *State and peasant in contemporary China*. Berkeley and Los Angeles: University of California Press.

OIM/INEC/CPSUDE (Organización Internacional para las Migraciones/Instituto Nacional de Estadísticas y Censos/Agencia Suiza para el Desarrollo y la Cooperación). 1999. Caracteristicas Socio-Demograficas de la Población Rural de Nicaragua, February.

OIM/INEC/UNFPA (Organización Internacional para las Migraciones/Instituto Nacional de Estadísticas y Censos/Agencia Suiza para el Desarrollo y la Cooperación/Fondo de Población de las Naciones Unidas). 1997. Migraciones internas en Nicaragua. Managua, August.

ONE (Oficina Nacional de Estadísticas). 1998. *Anuario estadístico de Cuba 1996*. Havana: ONE.

————. 2003. *Anuario estadístico de Cuba 2002*. Havana: ONE.

————. 2004. *Anuario estadístico de Cuba 2003*. Havana: ONE.

————. 2006. *Anuario estadístico de Cuba 2005*. Havana: ONE.

————. 2007. *Anuario estadístico de Cuba 2007. Series estadísticas*. Havana: ONE.

Pagés, Raisa. 2002. Mayor empuje para el sector agropecuario. *Granma Internacional*, February 10.

————. 2005. La ley que desató la guerra económica contra Cuba. *Granma*, May 7.

Pallot, Judith, and Tat'yana Nefedova. 2007. *Russia's unknown agriculture: Household production in post-communist Russia*. New York: Oxford University Press.

Paneque Brizuelas, Antonio. 1997. Posibles soluciones al problema agropecuario nacional? Por cuarta vez se reunen las UBPC para corregir el tiro. *Granma Internacional*, October 5.

Pastor, Manuel, Jr., and Andrew Zimbalist. 1995. Waiting for change: Adjustment and reform in Cuba. *World Development* 23 (5): 705–20.

Patel, Raj, and Maximilian Eisenburger. 2003. Agricultural liberalization in China: curbing the state and creating cheap labor. Policy brief no. 9. Oakland, CA: Food First.

Peña Castellanos, Lazaro. 2002. The sugar-cane complex: Problems of competitiveness and uncertainty in a crucial sector. In Monreal 2002a, 96–118.

Pérez Marín, Enrique, and Eduardo Muñoz Baños. 1992. Agricultura y alimentación en Cuba. *Agrociencia*, serie Socioeconómica 3 (2): 15–46.

Pérez Rojas, Niurka, and Cary Torres Vila. 1996. Las UBPC: Hacia un nuevo proyecto de participación. In Colectivo de Autores 1996, 46–67.

Pérez Villanueva, Omar Everleny. 2004a. The cuban economy today and its future challenges. In Domínguez, Pérez Villanueva, and Barberia 2004, 49–88.

————. 2004b. The role of foreign direct investment in economic development: The Cuban experience. In Domínguez, Pérez Villanueva, and Barberia 2004, 161–97.

Perry, Elizabeth, and Christine Wong, eds. 1985. *The political economy of reform in post-Mao China*. Cambridge: Council on East Asian Studies/Harvard University.

Pizarro, Roberto. 1987. The new economic policy: A necessary readjustment. In Spalding 1987, 217–32.

PNUD (Programa de las Naciones Unidas para el Desarrollo). 2002. *Informe sobre desarrollo humano 2002*. Mexico City: Ediciones Mundi-Prensa.

Polanyi, Karl. 1944. *The great transformation: The political and economic origins of our time*. Boston: Beacon Press.

Powell, Kathy. 2004. Mexican and Cuban agricultural policies: Political implications for small producers. Paper prepared for the XXV conference of LASA, Las Vegas, NV, October 7–9.

Putterman, L. 1988. Group farming and work incentives in collective-era China. *Modern China* 14 (4): 419–50.

Quintana Mendoza, Didio. 1996. Ingresos de la población por territorios en los '90. *Cuba: Investigación Económica* 2 (2): 1–13.

———. 1998. Evolución de los ingresos de la población en los años noventa: Diferenciación por territories. In *Cuba: Crisis, ajuste y situación social*, ed. Ángela Ferriol Muruaga, Alfredo González Gutiérrez, Didio Quintana Mendoza, and Victoria Pérez Isquierdo, 55–75. Havana: Editorial de Ciencias Sociales.

Rello, Fernando. 1996. Efectos sociales de la globalización sobre la economía campesina. Unpublished document. Mexico: CEPAL.

Renzi, María Rosa. 1996. Programas de estabilización y ajuste estructural en Nicaragua: Sus efectos económicos y sociales. Unpublished manuscript.

Reygadas, Luis. 2006. Latin America: Persistent inequality and recent transformations. In *Latin America after neoliberalism: Turning the tide in the twenty-first century*, ed. Eric Hershberg and Fred Rosen, 120–43. New York: New Press.

Riskin, Carl. 1987. *China's political economy: The quest for development since 1949*. Oxford: Oxford University Press.

Riskin, Carl, and Li Shi. 2001. Chinese rural poverty inside and outside the poor regions. In *China's retreat from equality*, ed. Carl Riskin, Zhao Renwei, and Li Shi, 329–44. Armonk, NY: M. E. Sharpe.

Robinson, William I., and Kent Norsworthy. 1984. *David and Goliath: The U.S. war against Nicaragua*. New York: Monthly Review Press.

Rodríguez, José Luis. 1998. Hay valores que la economía no puede medir e igualmente el precio que habría que pagar, si se pierden, es insoportable para un pueblo digno como el nuestro. Speech to the Asamblea Nacional del Poder Popular, December 21, 1998. Published in *Granma*, December 23.

Rodríguez Alas, Tomás Ernesto. 2002. Ajuste estructural y desarrollo rural en Nicaragua. *Cuadernos de Investigación* 16 (Nitlapan-UCA).

Rodríguez Eduarte, Yamila. 1997. El hilo de Ariadna de las UBPC. *Juventud Rebelde*, September 14.

Rosenberg, Jonathan. 1992. Cuba's free-market experiment: Los mercados libres campesinos, 1980–1986. *Latin American Research Review* 27:51–89.

Rosset, Peter. 1997. Cuba: Ethics, biological control, and crisis. *Agriculture and Human Values* 14:291–302.

Ruben, Ruerd, and Edoardo Masset. 2003. Land markets, risk and distress sales in Nicaragua: The impact of income shocks on rural differentiation. *Journal of Agrarian Change* 3 (4): 481–99.

Rylko, Dmitri, and Robert W. Jolly. 2005. Russia's new agricultural operators: Their emergence, growth, and impact. *Comparative Economic Studies* 47:115–26.

Sáez, Hector. 2003. A transition to sustainable agriculture in the 1990s? A case study of Santo Domingo, Cuba. Unpublished manuscript, University of Vermont.

Saldomando, Angel. 1992. *El retorno de la AID: El caso de Nicaragua—Condicionalidad y reestructuración conservadora.* Managua: Ediciones CRIES.

Sato, Hioshi. 2003. *The growth of market relations in post-reform rural China.* London: Routledge Curzon.

Sazanov, Sergei, and Damira Sazanova. 2005. Development of peasant farms in Central Russia. *Comparative Economic Studies* 47:101–14.

Scott, Christopher. 1996. The distributive impact of the new economic model in Chile. In Victor Bulmer-Thomas, 147–84.

Selden, Mark. 1988. *The political economy of Chinese socialism.* New York: M. E. Sharpe.

Serra, Luis, and Gerardo Castro. 1992. El crédito agrícola y los pequeños productores en Centroamérica. In *Alternativas campesinas: Modernización en el agro y movimiento campesino en Centroamérica,* ed. Klaus-D. Tangermann and Ivana Ríos Valdés, 209–41. Managua: Latino Editores.

Smith, Adam. 1946. *An inquiry into the nature and causes of the wealth of nations.* New York: Doubleday.

Somers, Margaret R. 2005. Beware trojan horses bearing social capital: How privatization turned *Solidarity* into a bowling team. In *The politics of method in the human sciences,* ed. George Steinmetz, 233–74. Durham: Duke University Press.

Spadoni, Paolo. 2004. Reasons behind Cuba's dollar ban. *Orlando Sentinel,* November 22.

———. 2008. Cuba's current economic situation: Macroeconomic performance, structural changes, and future challenges. Presented at the "Cuba: New Research Directions" Conference, University of California, Irvine, May 2–3.

Spalding, Rose J., ed. 1987. *The political economy of revolutionary Nicaragua.* Winchester: Allen and Unwin.

———. 1994. *Capitalists and revolution in Nicaragua: Opposition and accommodation, 1979–1993.* Chapel Hill: University of North Carolina Press.

———. 2007. Poverty politics in Nicaragua. Paper prepared for the XXVIII International LASA conference, Montreal, Canada, September 5–8.

SPAR (Sistema Público Agropecuario-Rural). 2008. Alimentos para el pueblo . . . Alimentos para la vida. Unpublished report.

Spoor, Max. 1994. Structural adjustment and grain markets in Nicaragua: From market substitution to market development? Paper presented at the XVIII LASA conference, Atlanta, GA, March 10–13.

———. 1995. *The state and domestic agricultural markets in Nicaragua: From interventionism to neoliberalism.* London: Macmillan.

———. 2002. Policy regimes and performance of the agricultural sector in Latin America and the Caribbean during the last three decades. *Journal of Agrarian Change* 2 (3): 381–400.

Spoor, Max, Nico Heerink, and Futian Qu, eds. 2007. *Dragons with clay feet? Transition, sustainable land use, and rural environment in China and Vietnam.* Lanham, MD: Lexington Books.

Stahler-Sholk, Richard. 1990. Stabilization policies under revolutionary transition: 1979–1990. Ph.D. diss., University of California, Berkeley.

———. 1996. Structural adjustment and resistance: The political economy of Nicaragua under Chamorro. In *The undermining of the Nicaraguan revolution,* ed. Gary Prevost and Harry Vanden, 74–113. London: Macmillan.

Stark, David. 1996. Recombinant property in East European capitalism. *American Journal of Sociology* 101 (4): 993–1027.

Stok, Gustavo. 2005. La brecha más profunda. *Americaeconómia* (May 20–June 9): 68–69.

Szelenyi, Ivan, and Eric Kostello. 1996. The market transition debate: Toward a synthesis? *American Journal of Sociology* 101 (4): 1082–96.

Thiesenhusen, William C. 1992. *Broken promises: Agrarian reform and the Latin American campesino.* Boulder, CO: Westview Press.

Thompson, Ginger. 2004. In reply to tightening of sanctions, Castro bans the Yankee dollar. *New York Times,* October 27.

Thrupp, Lori Ann. 1992. *Bittersweet harvests for global supermarkets: Challenges in Latin America's agricultural export boom.* Washington, D.C.: World Resources Institute.

Tiempos del Mundo. 2000. El occidente del país al borde de la emergencia económica. *Tiempos del Mundo,* July 6.

Togores González, Viviana. 2000. Cuba: Efectos sociales de la crisis y el ajuste económico de los 90s. Paper presented at the XXII conference of LASA, Miami, FL, March 16–18.

———. 2004. Ingresos monetarios de la población, cambios en la distribución y efectos sobre el nivel de vida. In *15 años del Centro de Estudios de la Economía Cubana,* ed. Colectivo de Autores, 112–35. Havana: Editorial "Felix Varela."

Torres Vila, Cary, and Niurka Pérez Rojas. 1996a. Apuntes sobre el problema de la fluctuación laboral en las UBPC. Colectivo de Autores 1996, 94–103.

———. 1996b. La apertura de los mercados agropecuarios en Cuba: Impacto y valoraciones. In Colectivo de Autores 1996, 177–206.

Travers, S. Lee. 1985. Getting rich through diligence: Peasant income after the reforms. In Perry and Wong 1985, 111–30.

Unger, Jonathan. 2002. *The transformation of rural China.* Armonk, NY: M. E. Sharpe.

UNO (Unión Nacional Opositora). 1989. Programa de Gobierno de la Unión Nacional Opositora. Mimeographed.

Utting, Peter. 1987. Domestic supply and food shortages. In Spalding 1987, 127–48.

———. 1991. *Economic adjustment under the Sandinistas: Policy reform, food security, and livelihood in Nicaragua.* Geneva: UNRISD.

Valdés Paz, Juan. 1992. *Procesos agrarios en Cuba, 1959–1995.* Havana: Editorial de Ciencias Sociales.

Velikii, P. P., and M. Iu. Morekhina. 2006. The adaptive potential of rural society. *Sociological Research* 45 (6): 51–67.

Veltmeyer, Henry, James Petras, and Steve Mieux. 1992. *Neoliberalism and class conflict in Latin America.* New York: St. Martin's Press.

Villegas Chádez, Rubén. 1996. Cerca del sentimiento de dueño en las Unidades Básicas de Producción Cooperativa (UBPC). In Colectivo de Autores 1996, 104–19.

von Braun, Joachim, Matin Qaim, and Harm tho Seeth. 2000. Poverty, subsistence production, and consumption of food in Russia: Policy implications. In *Russia's agro-food sector: Towards truly functioning markets,* ed. Peter Wehrheim, Klaus Frohberg, Eugenia Serova, and Joachim von Braun, 301–21. Boston: Clair Academic Publishers.

Walker, Kathy Le Mons. 2008. From covert to overt: Everyday peasant politics in China and the implications for transnational agrarian movements. *Journal of Agrarian Change* 8 (2/3): 462–88.

Weeks, John. 1995. Macroeconomic adjustment and Latin American agriculture since 1980. In *Structural adjustment and the agricultural sector in Latin America and the Caribbean*, ed. John Weeks, 69–91. London: St. Martin's Press.

———. 1992. Trade liberalization, market deregulation, and agricultural performance in Central America. *Journal of Development Studies* 35 (5): 48–75.

Wegren, Stephen K. 1998a. *Agriculture and the state in Soviet and post-Soviet Russia.* Pittsburgh: University of Pittsburgh Press.

———. 1998b. Russian agrarian reform and rural capitalism reconsidered. *Journal of Peasant Studies* 26 (1): 82–111.

———. 2004. Rural adaptation in Russia: Who responds and how do we measure it. *Journal of Agrarian Change* 4 (4): 553–78.

———. 2008. The limits of land reform in Russia. *Problems of Post-Communism* 55 (2): 14–24.

Wegren, Stephen K., Valeri V. Patsiorkovski, and David O'Brien. 2006. Beyond stratification: The emerging class structure in rural Russia. *Journal of Agrarian Change* 6 (3): 372–99.

Wells, Miriam J. 1996. *Strawberry fields.* Ithaca: Cornell University Press.

Williamson, John. 1990. *Latin American adjustment: How much has happened?* Washington, D.C.: Institute for International Economics.

Wright, Julia. 2005. *Falta Petroleo! Perspectives on the emergence of a more ecological farming and food system in post-crisis Cuba.* Ph.D. diss., University of Wageningen.

Xie Yu, and Emily Annum. 1996. Regional variation in earnings inequality in reform-era urban China. *American Journal of Sociology* 101 (4): 950–92.

Yao, Yang. 2007. The Chinese land tenure system: Practice and perspectives. In *The dragon and the elephant: Agricultural and rural reforms in China and India*, ed. Ashok Gulati and Shenggen Fan, 49–70. Baltimore: Johns Hopkins University Press.

Yardley, Jim. 2008. China enacts major land-use reform for farmers. *New York Times*, October 20.

Zavisca, Jane. 2003. Contesting capitalism at the post-Soviet dacha: The meaning of food cultivation for urban Russians. *Slavic Review* 62 (4): 786–810.

Zhong, Funing. 2007. The contribution of diversification to the growth and sustainability of Chinese agriculture. In Spoor, Heerink, and Qu 2007, 83–92.

Zhou, Kate Xiao. 1996. *How the farmers changed China: Power of the people.* Boulder, CO: Westview Press.

Rural Studies Series
Stephen G. Sapp, General Editor

The Estuary's Gift: An Atlantic Coast Cultural Biography
David Griffith

Sociology in Government: The Galpin-Taylor Years in the U.S. Department of Agriculture, 1919–1953
Olaf F. Larson and Julie N. Zimmerman
Assisted by Edward O. Moe

Challenges for Rural America in the Twenty-First Century
Edited by David L. Brown and Louis Swanson

A Taste of the Country: A Collection of Calvin Beale's Writings
Peter A. Morrison

Farming for Us All: Practical Agriculture and the Cultivation of Sustainability
Michael Mayerfeld Bell

Together at the Table: Sustainability and Sustenance in the American Agrifood System
Patricia Allen

Country Boys: Masculinity and Rural Life
Edited by Hugh Campbell, Michael Mayerfeld Bell, and Margaret Finney

Welfare Reform in Persistent Rural Poverty: Dreams, Disenchantments, and Diversity
Kathleen Pickering, Mark H. Harvey, Gene F. Summers, and David Mushinski

Daughters of the Mountain: Women Coal Miners in Central Appalachia
Suzanne E. Tallichet

American Guestworkers: Jamaicans and Mexicans in the U.S. Labor Market
David Griffith

The Fight Over Food: Producers, Consumers, and Activists Challenge the Global Food System
Edited by Wynne Wright and Gerad Middendorf

Stories of Globalization: Transnational Corporations, Resistance, and the State
Edited by Alessandro Bonanno and Douglas H. Constance